CONVERSATIONS
ON THE
DARK SECRETS
OF PHYSICS

CONVERSATIONS ON THE DARK SECRETS OF PHYSICS

Edward Teller
Wendy Teller
and
Wilson Talley

PLENUM PRESS • NEW YORK AND LONDON

Library of Congress Cataloging-in-Publication Data

Teller, Edward, 1908-
    Conversations on the dark secrets of physics / Edward Teller,
  Wendy Teller, and Wilson Talley.
        p.   cm.
    Includes bibliographical references and index.
    ISBN 0-306-43772-4
    1. Physics.   I. Teller, Wendy.   II. Talley, Wilson.   III. Title.
  QC21.2.T45 1991
  530--dc20   91-8626
                                                                    CIP

The poem "Perils of Modern Living" reprinted by permission;
© 1956, 1984 The New Yorker Magazine, Inc.

ISBN 0-306-43772-4

© 1991 Edward Teller
Plenum Press is a division of Plenum Publishing Corporation
233 Spring Street, New York, N.Y. 10013

Printed in the United States of America

This book is dedicated to the
Fannie and John Hertz Foundation

PREFACE

The idea for this book began over four decades ago when Edward Teller began teaching physics appreciation courses at the University of Chicago.

Then, as now, Dr. Teller believes that illiteracy in science is an increasingly great danger to American society, not only for our children but also for our growing adult population.

On one hand, the future of every individual on this globe is closely related to science and its applications. Fear of the results of science, which has become prevalent in much of the Western World, leads to mistaken decisions in important political affairs. But this book speaks of no fears and of no decisions—only of the facts that can prevent one of them and indirectly guide the others.

From the perspective of this book, a second point is even more significant. The first quarter of this century has seen the most wonderful and philosophically most important transformation in our thinking. The intellectual and aesthetic values of the points of view of Einstein and Bohr cannot be overestimated. Nor should they be hidden at the bottom of tons of mathematical rubble.

Our young people must be exposed to science both because it is useful and because it is fun. Both of these qualities should be taken at a truly high value.

Adults should be interested in science because it is a part of our cultural heritage and because the new technologies that are entering our society should be understood by as many of us as is possible.

It is our hope that this book will enable many otherwise-educated adults to catch up on the new physics so that they can properly contribute to the dialogue on the scientific and technological decisions that will shape our future. Also, we invite them to join us in an appreciation in the sheer joy of science.

The reader will find that equations are used in the text. Some writers avoid any and all equations, fearing that they will frighten off readers. We have deliberately included them to summarize the words in the text, and the lay reader need not be afraid to glance at them and even make a small attempt to decode them (the key to the code is always provided in the text). Like the sketches which also illustrate the words in this book, equations should be thought of as a form of summary.

To capture the essence of his lectures, Dr. Teller and his daughter, Wendy, began working on a manuscript. (As you will see, the footnotes in the text sometimes contain a dialogue between ET and WT.) They were joined in their effort by Wilson Talley (who also appears in the footnotes, joining the original WT).

The precipitating event that led to the completion of this book was an action by the Fannie and John Hertz Foundation. The Foundation, established by the founder of the Hertz Corporation and the Yellow Cab Company, began a series of experiments in undergraduate education, including students at primary and secondary schools. Among other projects, it was decided that Dr. Teller would be supported in teaching an updated "Physical Sciences Appreciation" course to high school students and teachers in the Livermore Valley area of California. The course was sponsored by the Foundation, the University of California, Davis/Livermore Department of Applied Science, and the Lawrence Livermore National Laboratory. We are indebted to those literally hundreds of students, as well as the thousands who have heard Dr. Teller speak on the appreciation of science over the past decades.

Along the way to completing this book, we owe a particular debt to several individuals. Paul Teller, Edward's son, read portions of the manuscript. Joanne Smith, Patty French, and Judy Shoolery took dictation, typed, and retyped various parts. Helen Talley, Wilson's wife, entered much of the original manuscript into the Macintosh and then gamely read subsequent versions for intelligibility. Because the "proof of concept" of the book was the course given at Livermore, we should credit Sue Anderson, Matt DiMercurio, Tom Harper, Barbara Nichols, Jaci Nissen, Maria Parish, Kathryn Smith, and Charlie Westbrook for their assistance in keeping that activity on line.

CONTENTS

PROLOGUE—A WARNING

> ". . . *Denn die Bücher ohne Formeln*
> *Haben meistens keinen Sinn . . .*"
> —From an apocryphal adaptation*
> of the *Three Penny Opera*

I will use mathematics because physics without mathematics is meaningless. Some readers don't know mathematics so I will try not to use mathematics without explaining it, and those readers who already know it will have to be patient and might even enjoy it, since I will try to explain in an unusual way. I want to warn you—I will say quite a few things that everybody understands and I will say a few things that nobody understands and even some things that nobody can understand. I take this liberty because it is an actual picture of what scientists do. If somebody follows everything I say (it may

---

* This adaption was written about 1932 by an obscure Hungarian poet for Max Born's fiftieth birthday.

happen) I will be very pleased. But I do not expect it, because the world is usually so put together that everyone runs into something he doesn't understand and experiences the limit of what he can understand. I would like to demonstrate that these limits exist.

I have one more philosophical (i.e., irrelevant) remark. It is often claimed that knowledge multiplies so rapidly that nobody can follow it. I believe that this is incorrect. At least in science it is not true. The main purpose of science is simplicity and as we understand more things, everything is becoming simpler. This, of course, goes contrary to what everybody accepts.

I will start by explaining Einstein which is considered the most complicated of tasks. Nobody can understand Einstein. An American soap advertisement claims its product is 99.44% pure. This, in America, is a very good standard. I claim that 99.44% of the western intellectuals have no idea what Einstein's theory is, what it means. I want you to join the remaining 0.56%.

I claim that relativity and the rest of modern physics is not complicated. It can be explained very simply. It is only unusual or, put another way, it is contrary to common sense.

The human mind is made in such a way that if I say something that you think is absurd the automatic reaction is that your earflaps come down and you stop listening. You should make an effort and continue to listen, remembering that I am going to say things that are "obviously" wrong; in fact, they are true.

Chapter 1

RELATIVITY

Space and Time of the Physicist

*In which a simple, absurd but correct proposal
of Einstein's is described which establishes
the framework for physics.*

I begin with the theorem of Pythagoras. As you probably know, Pythagoras was a Greek who lived in southern Italy. He was a philosopher, which, at that time, meant he was also a mathematician. He was a physicist. Unfortunately, he became involved in politics and therefore got into trouble. (In that, as in many other regards, some followed in his footsteps.)

The theorem of Pythagoras was known to the Babylonians a thousand years before Pythagoras, but to our knowledge Pythagoras was the first to prove it. The proof I will give is different from the one that Pythagoras found. It is also not precise, but it can be made precise if anybody is really interested in precision.

In Figure 1, we have a triangle with sides of length a, b, and c. The sides a and b form a right angle. Squares have been drawn on

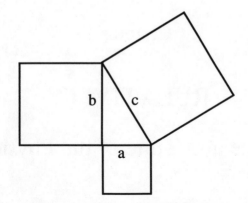

Figure 1. The Pythagorean theorem says that the sum of the squares of the legs of a right triangle is equal to the square of the hypotenuse.

each of the sides. The area of the square constructed on the side of length a is a^2 (a^2 means a times a). Similarly, the area of the square constructed on the side of length b is b^2 and the area of the square constructed on the side of length c is c^2. The theorem of Pythagoras says that $a^2 + b^2 = c^2$, that is, the sum of the areas of the two smaller squares is equal to the area of the big square.

To prove the theorem I draw two equal squares as in Figure 2. From each I will subtract four triangles, all equal in size but arranged differently. The four triangles are equal in area and the two big squares are equal in area, so the shaded area in the first square must be equal to the shaded area in the second square. Now the little square in Figure 2a has an area of a^2 and the larger square has an area of b^2. The shaded area in Figure 2b has an area of c^2. Thus we see that $a^2 + b^2 = c^2$.

The next statement, which we shall not prove, is, in a way, much more difficult, in a way much simpler. What is simple, what is difficult is different for different people.

As an introduction I want to draw in Figure 3 what is known as a Cartesian Coordinate system, named after the French philosopher Descartes. We have two perpendicular lines on a plane. Suppose

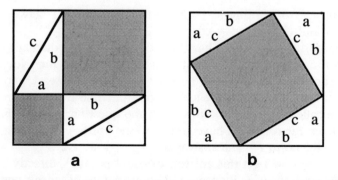

Figure 2. Copies of the triangle of Figure 1 can be rotated and flipped without changing the area. We can then rearrange these copies as in the two large squares to demonstrate that (a) the square of side a plus the square of side b will equal (b) the square of side c.

we have a point labeled P. If one starts at the intersection of the two lines, called the origin, one can reach P by moving a certain distance x along the horizontal line and then moving a certain distance y, parallel to the vertical axis. Then the two numbers x and y determine

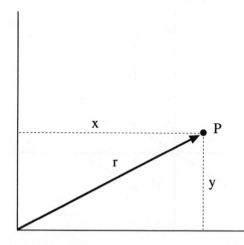

Figure 3. In the Cartesian coordinate system, the point P is reached by moving x units along the horizontal axis and y units up the vertical axis.

the point P. According to Pythagoras, the distance r between the point P and origin is $r^2 = x^2 + y^2$.

Unfortunately, space has three dimensions. If you want to fix a position in space and if you start at some "origin," then you have to say how far you go north, how far you have to go east, and how far you have to go up to reach the point. These three dimensions will be called x, y, and z. Now I ask the question: how far have I gone from the origin if I have gone x to the north, y to the east, and z up to the point P? The answer is $r^2 = x^2 + y^2 + z^2$.

To see how I get this answer, I look at point P', directly below P in Figure 4. By using Pythagoras, I know that the distance r' between P' and the origin is obtained from $(r')^2 = x^2 + y^2$. Now I consider the three points P, P', and the origin. If I connect these points with lines, they form a right-angled triangle and I can apply Pythagoras again, obtaining the answer

$$r^2 = (r')^2 + z^2 = x^2 + y^2 + z^2.$$

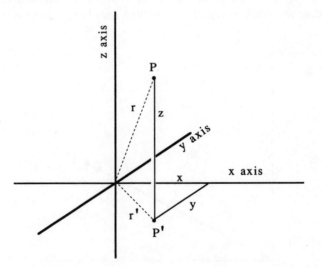

Figure 4.   The Pythagorean theorem allows us to find r, the distance from the origin to the point P, in three dimensions as well as two.

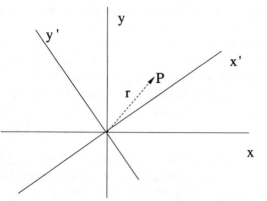

Figure 5.   Rotation of the axes x and y into x′ and y′, does not alter the distance from the origin to the point P; r is an invariant in this case.

So far I have dealt only with equations. Now I want to introduce an idea and this idea is an "invariant." An invariant is a quantity that does not change if you do certain things. For example, the distance between two points is an invariant under certain conditions. Consider the distance r in Figure 5. I could change the coordinate system. I will rotate the x and y lines (or axes, as they are called) and get the new x′ and y′ axes which are perpendicular to each other. Then I can get to the point P by going a distance x′ along the x′ axis and then going a distance y′ parallel to the y′ axis.* Then you can see that in the new coordinate system the numbers x′ and y′ which characterize P are different from x and y, but r remains the same. Therefore, I can say that I have an invariant $(x')^2 + (y')^2 = x^2 + y^2 = r^2$. No matter how I rotate my coordinate system, I get the same value of r, even though my values for x and y have changed.

---

* Please do not be confused that I call the axis and the distance one travels along the axis by the same name. This is what mathematicians do; they claim they are precise and then become completely imprecise. Physicists are worse, they claim they aren't precise and then, precisely when you aren't looking, they become precise.

Having discussed a little mathematics, we can start to talk about relativity. I now will discuss events, instead of points. In order to specify an event, I need four numbers: x, y, and z to specify the position and t to specify the time of the event. Four numbers are needed to describe each point and therefore we are discussing four dimensions. You may think that I am cheating, because time is very different from space. You will soon see that time is not all that different from space and this is the main point of Einstein's special relativity.

Let us start from the view that time and space are quite different. Suppose I am driving a car at 60 mph on a straight road. I push in my cigarette lighter and at the same instant I pass a hitchhiker. The cigarette lighter takes 15 seconds to pop out. I have two events; the first is my pressing the cigarette lighter in and the second is the cigarette lighter popping out. The hitchhiker will tell you (with few kind words for me) that the two events occurred 1/4 mile apart (since in 15 seconds, 1/4 of a minute, I have traveled 1/4 mile). I, on the other hand, will tell you that both events happened in the same place, about one foot from me, forward, and a little to the right. As far as I am concerned, I can say that the car is at rest and the world is moving backward.

The hitchhiker and I disagree on the distance between the two events. In this four dimensional world, in this geometry of space and time, r is no longer an invariant!

The circumstance that r is not an invariant was discussed very thoroughly several hundred years ago. This discussion is a part of what is called Galileo's principle, which says that the laws of physics are the same whether you describe the events as seen by an observer at rest or an observer in motion. But while the distance r is no longer an invariant, the time t, 15 seconds, that has passed between the two events is an invariant. The time is 15 seconds according to the watch that the hitchhiker is using and according to the watch that I am using. On that we all agree. It was true from the beginning, whenever that happened, up to the year 1905.

In the year 1905, the view that time is an invariant was changed by Einstein. This is the absurdity that I will discuss, namely, the time measured by me and the time measured by the hitchhiker are not the same. Einstein claims that the times don't agree, but he also says that there exists, instead, a different invariant.

Take two events. Suppose that t is the time between the two events, as measured by some observer, whether it is by me or the hitchhiker or some other observer who moves with respect to both me and the hitchhiker. We will call the speed of light c, it is 3×10^{10} cm/sec. Then ct is the distance that light can travel in the time between the two events, for instance, in 15 seconds. That is a big distance, a little more than a dozen times the distance to the moon. We shall, as before, call the observed distance between the two events r. Then we take the distance ct, square it, and subtract from it the square of the distance between the two events. We have then $(ct)^2 - (r)^2$.

In Einstein's theory, r is not an invariant, t is not an invariant, but $(ct)^2 - (r)^2$ is an invariant. This means that $(ct)^2 - (r)^2$ always has the same value, no matter whether I use my values for t and r or the hitchhiker's values for t and r or some other observer's values for t and r.

In the case we have been discussing from my viewpoint, $(ct)^2$ is very large (about twelve times the distance to the moon, squared) and r^2 is zero. From the hitchhiker's point of view, r^2 is $(1/4 \text{ mile})^2$, which is very small compared to my value for $(ct)^2$. Thus the difference between the time he observes and the time I observe is very small, so small that no one can measure it. So why all the fuss?

Let me jump to a case where Einstein's theory makes all the difference in the world. Let us say, for simplicity, that the moon is one light second away. (Actually, its distance is a little more than one light second.) That means that light takes just one second to go from the earth to the moon. Now I will send a light beam to the moon. I have two events: the first is the light beam leaving the earth, the second is the light beam arriving on the moon. I take the first event to be my initial point; the second event is 1 second later and 3×10^{10} cm away. That is, $c = 3 \times 10^{10}$ cm/sec, $t = 1$ sec, and $r = 3 \times 10^{10}$ cm, so $(ct)^2 = (3 \times 10^{10})^2$, $r^2 = (3 \times 10^{10})^2$, and $(ct)^2 - r^2 = (3 \times 10^{10})^2 - (3 \times 10^{10})^2 = 0$.

Since $(ct)^2 - r^2$ is an invariant, then for every observer this expression must be zero. Imagine an astronaut who leaves at the same time as the light beam and travels at 1/2 the speed of light.* For him the distance between the two events, which we will call r', will be smaller. The time, t', may also be different for him. But for him $(ct')^2 - (r')^2 = 0$. This means that $ct' = r'$. For our astronaut, the speed at which he sees light travel, $r'/t' = c$. We have our first absurd conclusion: the astronaut sees light traveling at the same speed as an observer on the earth. He ran after the light beam with 1/2 the speed of light. Common sense would suggest that he sees light moving ahead more slowly because he is running after it. However, if Einstein is right, light moves with the same velocity relative to *any* observer. No matter how fast you run after a light beam, it will gain on you and always gain with the same velocity. Thus the speed of light turns out to be an invariant.

The recognition that the velocity of light is the same for every observer was the outcome of an experiment designed to prove the opposite result. It is called the Michelson–Morley experiment. At the time of Michelson (up to 1887), people believed that light is a wave motion and that there is a substance in which these waves propagate, a substance that nobody had (or has) seen, called "ether." It was assumed then that we move relative to the ether and that this motion should express itself in an apparent change in the velocity of light. This change is what Michelson and Morley wanted to measure.

In order to measure the possible difference in light velocity, Michelson set up a device that split a light beam. Part of the light traveled along the direction of the earth's motion, the other part traveled perpendicular to the direction of the earth's motion.

To understand the problem, consider the analogous problem of two boys who are equally good swimmers who race in a river 1 km wide. Boy A will swim across the river and back. Boy B will swim upstream to a point 1 km up the river and then swim down-

---

* This will be excellent performance, even for an astronaut in the year 3000 AD.

stream 1 km to the starting point. Suppose that the river flows at 4 km/hr and the boys swim at 5 km/hr. Which boy will win the race?

Boy B, when he swims against the river, is actually progressing with a difference which is 1 km/hr and he will take one whole hour to get one km upstream. On the way back, he comes with (4 + 5) km/hr, that is, 9 km/hr. He makes the return trip in 1/9 hour or 6 2/3 minutes. Altogether, he takes 1 hr 6 2/3 min.

What about Boy A who swims across the river? How do you add the velocity of 5 km/hr that the boy swims upstream aslant to the 4 km/hr that is the current of the river? We draw the right triangle shown in Figure 6. The triangle is formed by the 5 km/hr velocity of the boy and the 4 km/hr velocity of the river and gives a net velocity of 3 km/hr perpendicular to the current, since by Pythagoras's theorem $3^2 + 4^2 = 5^2$. The boy's net velocity is 3 km/hr across the river. The river is only 1 km across and so he can swim across in 20 min and swim back in 20. He needs only 40 min and he wins the race easily.

In the actual case of the motion of the earth and propagation of light, the "boys" swim much faster and, by comparison, the "river of ether" is quite slow. So the difference in the race is small. But Michelson's sensitive apparatus could measure it.

Michelson expected that, when he was racing his light beams,

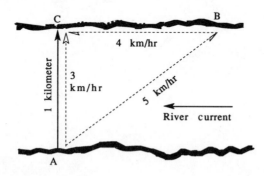

Figure 6.   To swim from point *A* to point *C*, the boy must swim upstream, as well as across the river.

the "across" light beam would win. This did not happen. The race was a precise tie. The experiment was repeated again and again, from 1881 to 1887. The more it was repeated, the more precise the tie became. Light was always moving with the same velocity. What would be the situation if we assumed that light is not a wave moving through ether?

Assume that instead of being a wave, light consists of particles. The light should move with light velocity relative to the source which emitted it—that is, the light should arrive at the same time whether it traveled against or across the moving ether. Thus, no surprise that the race was a tie. But this gave rise to more trouble.

We observe double stars which rotate around each other. We see them at a distance of many light years. We see these stars, if they are ten light years away, with a ten year delay. If we assume that light consists of particles, then the light from star B (in Figure 7), which is moving away from us, would arrive later than the light coming from star A. That is, the light would come toward us with light velocity minus the velocity with which star B is moving away from us, and the light from star A would come toward us at light velocity plus the velocity with which star A is moving toward us. Then the delay of light from star B would be more than the delay

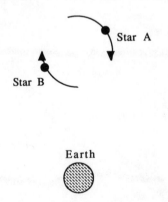

Figure 7.   Binary stars offer us an opportunity to verify that light velocity does not depend on the velocity of the body emitting it.

of light from star A, and the pattern of the motion of the double star, as we see it, would be very complicated.

But it isn't complicated. We see both stars with almost the same delay. Thus the idea that the velocity of light must depend on the velocity of the emitter cannot be correct. Now we have seen that light does not behave like a wave propagated in "ether" nor does it behave as we expect particles to do.

We are in trouble with either hypothesis as long as we stick to the usual description of space and time. But, according to Einstein, either the wave theory or the particle theory would work.* The velocity of light will appear the same to any observer, as long as $(ct)^2 - r^2$ is an invariant.

I have talked about a case where the invariant $(ct)^2 - r^2$ is positive. In fact, I have talked about another case, the light beam traveling to the moon, where the invariant was zero. There is a third case where the invariant is negative.

Consider two events which as far as you know are precisely simultaneous. The one occurs in Tel Aviv, the other occurs in New York. Both events occur at the same instant, say, 12:00 A.M. Tel Aviv time and 6:00 P.M. New York time. Then for these two events t will be zero, r will be 6,000 miles. Then $(ct)^2 - r^2$ will be negative. From this there follows a funny statement that Einstein himself made: Nobody can travel faster than light. Why?

If you believe that $(ct)^2 - r^2$ is an invariant, then I can show that it is impossible to travel faster than light. Assume you have two events that are not simultaneous. One happens a very short time after the other, but their distance is very great. Let us say the first event takes place on earth, the second event takes place on the moon, 1/10 sec later. What is the invariant? From our point of view, r is 3 $\times 10^{10}$ cm, t is 1/10 sec, $ct = 3 \times 10^9$. Therefore the invariant is negative. Assume somebody could travel from the earth to the moon in 1/10 sec. As far as he is concerned, r is zero so $(ct)^2 - r^2$ is positive. For us the invariant is negative, for him the invariant is positive and

---

* As we shall see in Chapter 10.

therefore the invariant is *not* an invariant. If you believe $(ct)^2 - r^2$ is an invariant, then it follows that one cannot travel faster than light.

You might say "this is all very nice mathematics, but why should I believe that $(ct)^2 - r^2$ is an invariant and that no one can travel faster than light?" We have made machines that accelerate particles to very high speeds. We find when we do this that, as the particles approach the light velocity, it gets harder and harder to accelerate them. You don't know yet, from this chapter, what energy is, but the fact is that we can give the particles more and more energy without any limit, but cannot give arbitrarily more velocity. We just approach the velocity of light. There are very accurate measurements proving Einstein's conclusion that no matter how you try to accelerate the particles, they won't go faster than light. This conclusion is verified.

So you say, "The conclusion is verified. I'll believe that, but why should I believe this invariant business? You haven't verified the invariant in a direct fashion." What scientists do is to make an assumption about how the world is put together, like the assumption that $(ct)^2 - r^2$ is the same for all observers. Then, if this assumption can explain verified facts, like the fact that light seems to travel at the same velocity for every observer or that particles cannot be accelerated to light velocity, the scientists accept the statement: "$(ct)^2 - r^2$ is an invariant" as a fact, at least until somebody comes along with a conclusion from this statement that seems to be contrary to what actually happens in the world. So the best we can do is to say: "Accept the 'fact' that $(ct)^2 - r^2$ is the same for all observers, because it leads to conclusions that agree with observations in the real world."

I have difficulty drawing in three dimensions and even more difficulty drawing four dimensions. So I will draw only two dimensions in Figure 8, the t dimension and the x dimension. Consider the two points P and Q. These two points occur at the same time $t = 0$. Their separation is x. The invariant for these two points will be negative. The two points P and Q will be called spacelike because there is a spatial difference between them.

On the other hand, P' and Q' occur at the same place, but there is a time difference between them. The invariant for P' and Q' will be positive and the two points will be called timelike.

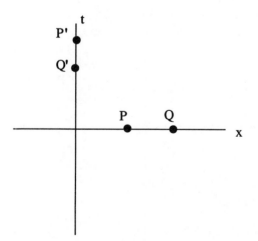

Figure 8. The points *P* and *Q* are "spacelike," because they occur at two different points in space; points *P'* and *Q'* are "timelike," because they happen at different times.

In Figure 9 I have changed the *t* axis to the *ct* axis, that is, I plot on the vertical coordinate *ct*. I have drawn the line $x = ct$ in Figure 9. Along this line, $x = ct$, the invariant is zero. The invariant is also zero along the line $x = -ct$. These lines, $x = ct$ and $x = -ct$ form what is known as the light cone.*

As you know, a cone is a point and straight lines radiating from this point in such a way as to include a fixed angle with one direction. I have not really drawn a cone, but I could have gone in the y direction or the z direction or any combination of x, y, and z directions; therefore, I am really talking about all the points or events which light could have reached in any direction. These points form the light cone. I will define regions, as pictured in Figure 9, as the future and the past. The points outside the light cone I will call the present, not just the points on the line $t = 0$, which was the way people thought

---

* In three dimensions (x, y, and ct) it would be a common cone. In four dimensions (x, y, z, and ct) it is still called a cone. In the two dimensions of Figure 9, the cone becomes two lines.

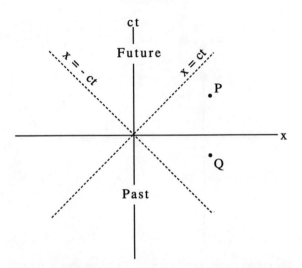

Figure 9. If we had included the y and z axes, the dashed lines would have formed a cone, the "light cone." All events within the upper cone are the future; all those within the lower cone are the past; all those outside the light cone, such as *P* and *Q*, are in the present.

before Einstein's revolution. Why do we call points that are outside the light cone the present? Clearly a point like *P* has a positive *t* coordinate and seems to be in my future. However, one can show that as long as the invariant is negative, that is, as long as *r* is greater than *ct*, there will be an observer from whose point of view *t* will be zero. From his point of view *P* will be simultaneous with his present. He must move with respect to you with a velocity less than *c* to accomplish this.

This is one of Einstein's most famous statements. It is the first real statement of his original paper. He criticized the idea of simultaneity. He called his theory "relativity" and his prime example was that the statement "two events are simultaneous" is true relative to one observer, but may not be so relative to another observer. I have oversimplified the situation. Consider the points *P* and *Q* in Figure 9. They are spacelike with respect to me, that is, with respect to the origin. With respect to each other they are timelike. Thus although the events P and Q are in my present, they may not be in the present of each other. Similarly, if I have two timelike events from my point of view, they may be simultaneous to each other.

Space and time are different in this sense, that the two timelike regions can at least be clearly separated into the future and the past. The spacelike regions form one continuous body and they cannot be naturally separated. We live in an assembly of stars, a hundred billion suns we call the Milky Way galaxy. There is another system of this kind known as the Andromeda Nebula. For the sake of simplicity, assume that the Andromeda Nebula is two million light years away; I always had the ambition to go there. (Actually, it is 1.8 million light years away, but this would make my numbers too complicated.) I want to go to Andromeda but, unfortunately, Einstein said that no one can move faster than light moves. My doctor says that it is unlikely that I will live two million years, so I cannot get there. Too bad for me.

But according to Einstein, I can get there in spite of what I have said. All I need to understand is the invariance of $(ct)^2 - r^2$. I will take off in a spaceship and I will assume that, though I cannot go faster than light, I can go almost as fast as light. The engineers tell me that this cannot be done and the engineers are right—for now; maybe we will invent something yet. At any rate, engineering is engineering and we are discussing physics. Let me assume that I can go almost as fast as light and I will take just a shade longer than two million years to get there. That is, you, who stayed at home, believe that I took a little more than two million years to get there and that I have gone a distance of two million light years. This means that ct will be a little bigger than r, but if I am very close to the velocity of light, then ct will only be a very little bit bigger than r. Therefore, the invariant will be positive, but small.

How do I feel about it? I will say that my departure and arrival are two events that occur in the same place, namely, inside the spaceship. Thus r is zero. The invariant must be the same and therefore must be small. The ct must be small and t must be small. From my point of view, the time that has passed is small. I can make it!

Now comes the really interesting part. I go to Andromeda and look around, take some scientific notes, turn around and, again with very great speed, come back. Let us assume that I do all this in one year flat, my time. I hope that there will be a big reception in New York. I will be a hero for having been to Andromeda and having come back.

I shall be disappointed, because I come back and here on earth four million years have passed, two million while I was going and two million returning. All of you shall be dead. Anybody who talks as I do, Hungarian, German, or English, shall be dead. The human race will have evolved into something entirely horrible, but something they imagine to be better. I believe they will be very tolerant. They will understand me, I will not understand them; they will be interested in me and very kindly and gently put me in a zoo.

This is all clearly nonsense. Does the time actually pass more slowly for me? This is a complicated question. As I leave earth and get further and further away, I get signals from earth with greater and greater delay. I make corrections for these delays according to my own watch. When we compare time, I already take into account not what I as the astronaut see, but what I have obtained from the earth and how I have corrected it. It is this complicated situation that then gives the final comparison when, with mission accomplished, I arrive back home.

If all these corrections are made and if you from the earth observe, you will see that my watch moves very slowly; you will see that my heart beats very slowly. For me, it is the real time, but for you, who look at me from the earth, you have to say: "How funny, his movements are slow, his heart beats slowly, his watch has almost stopped!"

As I move away from the earth very fast, I see the people on the earth move away from me very fast. While you see that my heart has slowed down very greatly, I should see that your hearts have slowed down just as much. Why is it, then, that in the end result you have aged much more than I? Where is the asymmetry in our situations? Why is it that when I come back and we can compare our clocks directly, it is I who have stayed young and you who have progressed in time four million years?

There are three occasions for lack of symmetry. While I travel at a uniform speed, there is no lack of symmetry, but when there is an acceleration, when I start to have a high speed, when I turn around in Andromeda, and when I stop at the end of the trip, these are the opportunities for asymmetries.

Now, two of these don't count: when I start and when I stop. At those times we can compare clocks without any corrections; there can be nothing wrong. But when I turn around in Andromeda, I have a big acceleration and that is a physical effect, something I experience and you back home do not. At that point we cannot compare watches directly, only distantly and we must apply correction factors. That is where the effective asymmetry comes in. This situation, together with some other ideas that Einstein put forward later, are the elements from which Einstein's magic created the theory of gravitation. I will return to this in a later lecture. For now it suffices to state that, while I turn around in Andromeda, the earth has jumped ahead of me four million years (from my point of view).

There is one more point that should be mentioned while we are discussing relativity. We have particles coming from outer space with very high energies and velocities very close to light velocity. These particles that arrive are stable particles, protons. As they hit particles in the air, nitrogen nuclei, for instance, they generate other particles, like mesons (called π mesons) which are unstable. These live only 10^{-8} seconds, that is, 10 times a billionth of a second. Then they turn into more stable mesons (called μ-mesons) with a lifetime of 2×10^{-6} seconds, that is, two millionths of a second. This life span has been measured in the laboratory. In two millionths of a second, a particle moving with approximately light velocity—3×10^{10} cm/sec—could travel 6×10^4 cm or 0.6 km, yet these particles generated 10 km over our heads arrive down here. How do they do it? They should have vanished after a relatively short distance. The explanation for the mesons' apparently long life is time extension. If you move along with the particle, it lives 2×10^{-6} seconds. If you view the particle from the earth, its lifetime is extended—like my life time was in my travel to Andromeda. This is the most direct proof of extension in time predicted by relativity.

Now I can stop with good conscience. I have proved relativity, or at least I have indicated a way such a proof is usually given.

But have I convinced you? Do you understand? Relativity is unconventional—is it also absurd?

To understand means (in functional terms) to know your way around. In this sense, you do not understand, if you are a novice. Perhaps you are on your way to understanding.

QUESTIONS

1. The proof in the text of the Pythagorean Theorem is incomplete. Improve it.

2. To make life easy on an astronaut, his craft accelerates at just one g until his velocity is approximately that of light. How long will this take?

3. In the Michelson–Morley experiment, the river's current is analogous to the speed at which the earth moves, which is approximately 3×10^6 cm/sec. The speed of the swimmer is analogous to the speed of light which is 3×10^{10} cm/sec. If the distance of the race is one meter, calculate who wins the race and by how much.

Chapter 2

STATICS

The Science of No Motion

*In which Archimedes takes a bath and thinks
of buoyancy, vectors, and other concepts
for which the Greeks, then, had no words.*

In Chapter 1 we did not discuss physics. Instead, we discussed the framework into which physics seems to fit in a most appropriate way. The beginnings of what we now call physics can be easily traced back to the Greeks, although they were not interested in the erratic and "unnatural" behavior of matter in motion. Their interest was in statics.

Archimedes lived 200 years before Christ, which is quite a jump backward from Einstein. I discussed Einstein first because he clarified the geometry of space and time. In many things that I'm going to write later, I will have to refer back to him because he has made physics simpler.* You will see as we proceed that the laws of conservation of energy and the conservation of momentum become a single law if you look at them from the point of view of Einstein. The laws of electricity and the laws of magnetism become unified and simpler if you look at them from the point of view of Einstein. Basically, everything I'm going to say will become simpler if we understand Einstein.

While relativity applies to almost all parts of physics it is, in a sense, not physics. It belongs to the mathematical (or perhaps geometrical) framework into which physics fits. The most elementary part of true physics may well be statics, the manner in which forces balance. This is the field in which the Greeks were interested and in which the greatest Greek mathematician and scientist, Archimedes, made his permanent contribution to physics.

There is a good story about Archimedes, which I recently discovered was not true, but that will not prevent me from repeating it. It is true that Archimedes lived in the Greek town on Sicily called Syracuse. The ruler, Heron, was a friend of Archimedes. Heron hired a goldsmith to make a crown for him and when the crown was finished, Heron suspected that the crown was made of gilded silver, instead of gold. Heron would have beheaded the goldsmith if the crown was not gold, but he wanted to be sure whether or not he was cheated. He asked Archimedes to help him, but he would not allow the crown to be harmed. He demanded what is now called "nondestructive testing."

To think about the problem, Archimedes decided to take a bath, since he did his best thinking in the bath.† Archimedes filled the bath

---

* WT: You will see "simple" is a key word in this book. It is my belief that there must be a word in Hungarian which has a meaning which is a combination of simple, elegant, and esthetically pleasing. Since ET does not speak English, we must forgive his not so simple interpretation of the word "simple."

ET: The Hungarian word for "simple" is "egyszerü." Its literal meaning is "like one." It seems that in Hungarian simple is the same as unifying. The German word is "einfach." The English word, "simple," I am told, also has the connotation, "not too bright."

† ET states that, in this regard, he is a follower of Archimedes.

to the brim, got in, and you won't be surprised to hear that the water spilled. At this point, Archimedes jumped out of the bath and ran through the streets of Syracuse naked, yelling, "Eureka! I have found it!"

You may wonder what Archimedes had found. Perhaps, you say, he wanted a good excuse to split from the house before his wife found the mess in the bathroom. Actually, he could measure the weight of the water that the crown displaced. He knew the density of the water, so he could find the volume of the displaced water, which was equal to the volume of the crown. Since he could also measure the weight of the crown, he could divide this weight by the volume of the crown to find its density. Gold and silver have different densities, so he could find out whether the crown was made of gold or silver.

I cannot tell whether the goldsmith lost his head, because I do not know what Archimedes found (and, besides, the story is not true).

There is another solution to the problem which is more instructive and which is a proper part of statics. Undoubtedly, Archimedes was not the first to use this method, but he thought about it more clearly than anybody before him. The solution involved is buoyancy and the solution is called Archimedes's principle. Take a crown and immerse it in water. Attach a string to the crown and measure its apparent weight while immersed. The pull on the string is not the same as in the absence of water. The water presses on the crown from all directions and, if you add up all the forces of the water pressure, you get the buoyancy which makes the crown appear lighter. Archimedes's principle is that the crown will lose as much weight as the weight of the water it displaces. Now gold has a density 18 times the density of water and the weight of the immersed crown should be 17/18 of the weight of the free crown.

To prove Archimedes's principle, imagine that the water that was displaced by the crown is put back into its original location. Now if you could attach a string to this "watery crown," it would not pull on the string. If, instead of the "watery crown," we put in a different solid having the same density as the water, the pull will still be zero. Since the buoyancy depends only on the pressure, the "watery crown" should lose just as much weight as the gold crown.

Archimedes also worked with other inventions. He actually helped Syracuse in its military defense against the Romans. With his own hands, Archimedes managed to lift a big ship. He did this with levers and pulleys, but before we discuss levers, we must discuss vectors and forces.

The concept of a vector is something that Archimedes did not deal with, but it will be very important in understanding forces. A vector is usually defined as something that has a magnitude and a direction. That is a poor definition. I would like to give an example of a vector, namely a displacement. A displacement is characterized by two points in space, A and B, and an arrow from A to B (Figure 1). I add that all displacements are equal, no matter where they start, as long as they are parallel (point in the same direction) and have the same length.

Suppose I have two displacements, **a** and **b** (vectors are denoted by writing them in boldface). How can I add them? I will add them by transporting the start of **b** to the end of **a**. Then the sum of the two vectors will be the vector **c** which starts at the start of **a** and ends at the end of **b**, as in Figure 2.

Would it make any difference if I started with **b** and then added **a**? It would not make any difference. Draw the addition of **a** + **b**. If

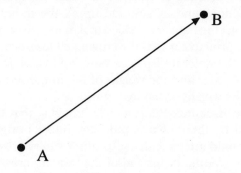

Figure 1. The arrow represents a displacement from the point A to the point B.

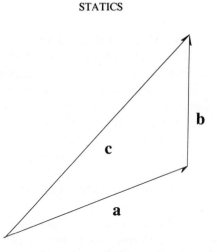

Figure 2.   The vector **c** is formed by adding the displacements **a** and **b**; **c** = **a** + **b**.

we rotate the figure by 180° around the midpoint of **c**, we get the addition of **b** + **a** = **c** with a change in the direction of the arrows (hence the "−" signs), as in Figure 3. This law of addition of displacements, that **b** + **a** = **a** + **b**, is called the parallelogram law, because a figure showing **b** + **a** and **a** + **b** forms a parallelogram.

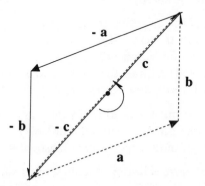

Figure 3.   The law that **c** = **a** + **b** = **b** + **a** is called the "parallelogram law," because the four displacements **a**, **b**, −**a**, and −**b** form a parallelogram.

A vector is a generalization of the concept of a displacement. (Generalization is a bad habit that physicists have picked up from mathematicians.) Physicists will tell you that forces are vectors. What they mean is that forces can be added (among other things) the same way in which displacements can be added.

Now a force is something that pushes or pulls in some direction with some strength. This definition sounds similar to the definition of a displacement, so it is not so surprising that forces are added together the same way displacements are. Still, it would be nice to have some sort of proof that the addition must be performed in the same way.

I will not prove that forces add like displacements, I will only indicate that all objects sharing some simple properties (like displacements and vectors) must add in the same way. I will do this in a very dirty manner, namely, by adding assumptions as I need them, but I will demand that all the assumptions be reasonable. This, you may say, is like the cook who decides to make dinner and, every time she* realizes she needs something, runs to the store to get it. It is not a very efficient way to cook dinner, or to prove a theorem, but at least the cook understands why she will need each ingredient and hopefully we will understand why we need each assumption.

I must start by introducing my first axioms. (They are the ones the cook finds on the shelf.)

1. Two equal forces acting in the same direction shall add up to a force in the same direction with twice the magnitude. (Similarly, a force and 1/2 a force in the same direction will add to 1 + 1/2 times the force in the same direction and similarly for any other numbers.)

The resultant force must indeed be in the same direction by symmetry. In what other direction should it be? We are simply assuming it is twice the magnitude, which seems reasonable.

2. Two equal forces in opposite directions shall cancel.

---

* WT: ET sometimes forgets that men can cook and that women can be mathematicians. However, he means no harm.

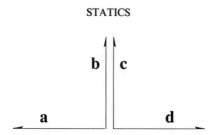

Figure 4. Given four unit forces, **a** and **d** pointing left and right, respectively, and **b** and **c** pointing up, we are interested in finding the single force that can replace all of them.

I can rotate the configuration of the two forces by 180°. This gives exactly the same configuration. Note that the only force that equals itself after rotation of 180° is the zero force. Symmetry demands that the forces cancel. It is like the medieval ass who is exactly between the two haystacks. He cannot decide which haystack to go to, so he just stays put and dies of hunger.

3. Whatever pattern the forces and their sum establish will remain valid if I rotate the whole thing.
4. It does not matter in which order I add forces. In fact, they are supposed to act simultaneously.

Now I ask this question: What is the sum of four forces, each having unit length, two pointing up, one points to the right, and the last points to the left? (See Figure 4.) The answer follows immediately from the postulates. The horizontal forces cancel and the vertical forces add to a vertical force of length two. So what is so interesting in all of this?

The interest comes if I add the forces in a different order. Now I shall appear like the scatterbrain cook, adding postulates as I need them. First, I want to add **a** and **b**, which are two forces of equal strength. In which direction will the resultant force be?

5. The resultant force of the addition of two equal forces will have the direction of the internal* bisector of the added forces.

---

* The word "internal" means the obvious direction, rather than the opposite direction.

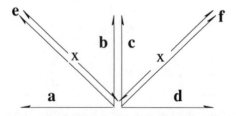

Figure 5.   Forces **a** and **b** add up to produce a single resultant force **e**, which points at 45° between the two. Forces **c** and **d** also produce a force **f** which is at right angles to the other resultant.

This is a reasonable assumption because of our old friend symmetry. So we know that the resultant force of **a** and **b** (which are at right angles to each other, see Figure 5) will be at a 45° angle and have some magnitude x. Similarly, I can add **c** and **d** to get a force at a 45° angle with the same magnitude. How do I know the magnitude x? For that purpose, I add the resultant vectors of magnitude x, which you will notice are at an angle of 90°. To do this, I need another postulate.

   6. If we know that the addition of the forces **a** and **b** give the resultant force **c**, then if we multiply **a** and **b** by a constant x, the resultant force of x**a** and x**b** will be x**c**, **c** multiplied by the same constant.*

In the first step we added **a** and **b** both of unit length to get a vector **e** of length x. Now we add **e** and **f** (both at right angles to each other and both of magnitude x) we know that the result must have magnitude $x \cdot x$ or x^2 and bisect the angle between **e** and **f**, which is in the direction "up." We know that the final force resulting from the addition of the four forces **a**, **b**, **c**, and **d** has magnitude 2, so we know that $x^2 = 2$. This means that $x = \sqrt{2}$, the number which, when multiplied by itself, gives 2.

---

* A very clever cook could derive this "postulate" from the one that vectors may be added in any order.

What we have proved by all this maneuvering is that two equal forces which are perpendicular to each other add to a force of magnitude $\sqrt{2}$ times the magnitude of the original forces. This is what is expected if forces are vectors and behave like displacements (to which the statement of Pythagoras $a^2 + b^2 = c^2$ applies). It is far from proving that forces are vectors. If you are not satisfied, I can assure you that with enough postulates (all of which make good sense) and patience, you can prove the parallelogram law for forces. If you do this, I must warn you, you are probably a terminal case—you will become a mathematician.

After this detour, we can return to Archimedes and his lever. A lever is basically a stick and a fixed point which we call a fulcrum. Archimedes was very proud of his lever and claimed that, given a fixed point, he could move the earth. His pride in the lever seems a bit unwarranted, since monkeys have been using sticks to pry things loose for millions of years. But it is not proven that they fully understand what they are doing.*

To make things simple, we will assume that there is no friction and the lever has no weight. I will now put two equal forces at equal distances from the fulcrum, as in Figure 6.

The lever will not move because of symmetry. Now what happens if I make one arm twice as long as the other and hang twice the weight on the short arm as I do on the long arm? We now have Figure 7.

I claim that the lever will again be in equilibrium and I will prove it by adding other "imaginary" forces shown as dotted arrows, as in Figure 8.

I imagine that I have a two unit force pointing down at D, which is the same distance from B as is A. Then to balance this new force, I will add a unit force pointing up at B and at C. For reasons of symmetry, these forces cannot turn the lever, and they will not displace it up or down because their sum is zero. The result is that the downward force at C is balanced and the forces at A and D cannot

---

* WT: One might ask Koko, the gorilla, for an explanation.

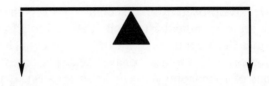

Figure 6. Two equal forces at equal distance from the fulcrum cancel each other out; the lever will not rotate.

turn the stick, for reasons of symmetry, and cannot move it because point *B*, the fulcrum, is fixed. Since the imaginary forces balance and the imaginary plus the original forces balance, the original forces must balance all by themselves. The result is that twice the force at unit distance balances with a unit force at twice the distance; that is, if the ratio of the arm lengths is 1 to 2, and the ratio of the forces is 2 to 1, we are in equilibrium.

Now I could go ahead and prove a similar statement for any ratio by using similar methods (but you will prefer to just take my word for it). If Archimedes wanted to move a very big ship, he had only to use a very long arm.

You might have the feeling that Archimedes is cheating, that he is getting something big (moving the ship) for something small (the force he exerts on the very long arm). To show you that Ar-

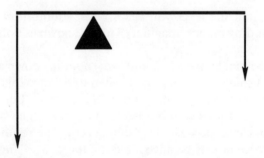

Figure 7. If the lever arms are not equal and the weights are also not equal, it is still possible to have a state of equilibrium where the lever will not rotate.

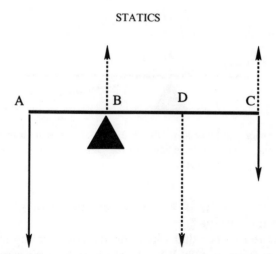

Figure 8.   The distances *AB*, *BD*, and *DC* are all the same. At the points *B*, *D*, and *C* we add fictitious forces (the dotted arrows) such that their vector sum is zero and that they produce zero torque.

chimedes is not cheating, I must introduce a new idea, the idea of energy. Actually, for the moment I will discuss only a special kind of energy, potential energy, the energy due to the position of a body in the gravitational field of the earth. Archimedes is not cheating and we can prove him an honest man by showing that the potential energy is conserved. Let us return to the case where one arm is twice the length of the other, and suppose that I have done the lifting. The lever has rotated and we get two similar triangles, the triangle on the right being twice as big as the one on the left (see Figure 9). (Actually, we don't have triangles because AA' and CC'' are arcs, not straight lines, but if the displacement is small, I can approximate arcs by straight lines.)

But the distance AA' is 1/2 the distance CC'. To move a unit object a unit distance, I must exert 1/2 the force over twice the distance. What we have conserved is the force times the distance. Force times distance is work, or change in potential energy and the total change in potential energy is zero. It is a very important idea

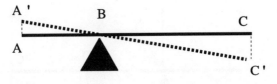

Figure 9. Twice the lever arm ($BC = 2 \cdot AB$) allows you to move twice the weight, but the work you do is equal to the gain in energy of the weight as it rises from point A to point A'.

in physics that energy is conserved. In this sense, you cannot get something for nothing.*

Please remember that force and distance (or rather force and displacement) are vectors. What I have actually done is multiply two vectors and we indicate this by $\mathbf{f} \cdot \mathbf{d}$, where $\mathbf{f}$ is the force and $\mathbf{d}$ is the displacement and the dot represents the dot product, which peculiarly enough is not a vector but just a scalar, in this case, an energy which has a magnitude but not a direction.

The simple multiplication which I had used did suffice in the cases where the force and the displacement were parallel. Suppose I have force as in Figure 10 which is not straight down, but also goes toward the fulcrum.

Now this new force can be written as the sum of two forces, one down and one toward the fulcrum, as indicated by the dotted arrows. (The horizontal arrow should really coincide with the lever.) The lever will be in equilibrium only if the part of the force pointing down is equal to the force on the other arm, providing, of course, that the two arms are equal. The part of the force pointing inward will push the lever, but since the lever is rigid and the fulcrum is fixed, it will have no effect on the lever.

Now to find the work, force times displacement, I should not use the length of the force, but only the length of the force projected onto the direction of the displacement. Take the perpendicular

---

* In physics, we look for quantities that are conserved. If we find them, we have made a step toward greater simplicity.

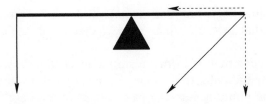

Figure 10.   Forces that do not act at right angles to the lever can be resolved into components along the lever and at a right angle to the lever.

shadow of **f** on **d**, and then multiply this by the length of **d**. If you know your trigonometry, you know that this is the same as multiplying the product of the magnitudes of **f** and **d** with the cosine of the angle between **f** and **d**. Therefore, our quantitative result is: the dot product is the product of the length of the two vectors times the cosine their directions include.

There is one more illustration of how vectors can be used that I want to present. That is the idea of torque. To illustrate this, I return to the lever with equal weights hung on equal arms. Each weight wants to turn the lever. The weight on the right wants to turn the lever clockwise and the weight on the left wants to turn the lever counterclockwise. If we use the law of levers mentioned above, this want to turn is proportional to the force and to the arm of the lever. This want to turn is called the torque.

If I now consider a force on the arm which can be in any direction, the only effective part of this force will be the component of the force which is in the plane perpendicular to the arm. The portion of the force which is along the arm does not know whether to rotate the lever in one direction or the other, so it has no effect. We are now taking a product of vectors by first projecting the force on a plane perpendicular to the arm **d** and then multiplying by the arm. If you know trigonometry, you will realize that this is just **f** times **d** times the sine of the angle between **f** and **d**. This kind of product, written as **f** × **d**, is called a cross product. The cross product **f** × **d** has a magnitude but is not described by that magnitude alone. One also has to add around which axis the turning is to occur. So we have a quantity which has a magnitude and a direction. The direction is that axis. You may suspect that $\mathbf{f} \times \mathbf{d}$, the torque, is a vector.

This means that, if I draw a simple lever, the direction of the torque is perpendicular to the plane of the page. But is the direction of the torque pointing into the page or out of the page? We know that the weight on the right side cancels the weight on the left side. We must make these two torques opposite in direction in order for torques to add like vectors.

In order to determine the direction of the torque $\mathbf{f} \times \mathbf{d}$, there is a simple convention: If I point my right thumb in the direction of $\mathbf{f}$, my index finger along the direction of $\mathbf{d}$, then my middle finger will point in the direction of the torque. After playing with this convention, you will see that the two torques on the simple lever cancel. You can prove to yourself that all the basic postulates that we made about forces hold, for example: two torques in the same direction add up to a torque in the same direction, also a pattern which is a valid torque addition remains a valid torque addition if I rotate the whole object in space; furthermore, the sum of two equal torques will give a new torque in the same plane that bisects the angle between the torques. If you believe these things about torques, then you can convince yourself that torques add up like vectors and actually are vectors.

Torques are different kinds of vectors from displacements. If I take a displacement and reflect it around a center, I multiply it by -1 and I obtain a displacement in the opposite direction.

What happens when I reflect the components of a torque around a center? F is multiplied by -1 and I obtain a force in the opposite direction (see Figure 11). I reflect $\mathbf{r}$, multiply it by -1, and obtain an arm in the opposite direction. The result is a torque which wants to turn in the same direction as the original torque. Reflection around a center is called an inversion. A vector which changes sign upon inversion, like a displacement, is called a polar vector. A vector which does not change sign upon inversion is called an axial vector.

Practically everything I have discussed in this chapter was known to Archimedes. The Greeks knew statics, but they did not seem to

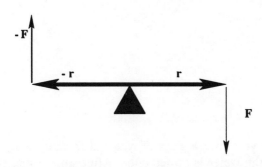

Figure 11. A torque is an example of an "axial vector" as it does not change when all vectors that produce it are inverted. Note that force does change if it is inverted; it is a "polar vector."

turn their interest to the laws of motion. Archimedes did not overcome this limitation.

There is one exception. Motion of the heavenly bodies did interest the Greeks. They invented explanations for these motions that happened to describe the facts, but did not explain them. It took the intellectual revival, the Renaissance, to provide a focus on the laws of motion. In this process, the explanation of the planetary system did play an essential role.

You may have accepted the discussion of statics more readily than our previous introduction to relativity. This is usual. We have all used levers to pry things loose or have seen pans of water overflow; few of us have ever worried about invariants. I want to tell you, however, that the mathematics and mental gymnastics needed to *properly* understand what Archimedes has done is actually *slightly* more difficult than that needed to understand Einstein. We don't fear the familiar and can, therefore, often comprehend it easily.

QUESTIONS

1. Using the postulates about the addition of forces in this chapter, find the force resulting from the addition of two equal forces at a 120° angle.

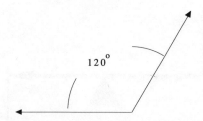

2. Suppose you have a glass of water at freezing temperature and floating in the water is an ice cube. After some time, the ice cube melts. Will the level of the water in the glass go up, go down, or stay the same?

3. A perpetual motion machine is designed as follows. Take a triangular block of wood as shown in the figure below and hang a heavy chain around it as shown. Then the weight of the chain on the long side, *AB*, is larger than the weight of the chain on the side *AC*. Thus the chain will start moving, producing energy for nothing. This contradicts the conservation of energy. Where is the fatal flaw in this machine?

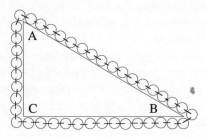

Chapter 3

A REVOLUTION IGNORED,
A REVOLUTION REPRESSED

*The story of the heliocentric system
for which one Greek had a good word,
but the other Greeks would not listen.*

If there ever was a misnomer, it is "exact science." Science has always been full of mistakes. The present day is no exception. And our mistakes are good mistakes; they require a genius to correct them. Of course, we do not see our own mistakes.

In this chapter, I will discuss the mistakes of Aristotle and how these mistakes were finally corrected two millennia later. This story is a lesson of which everybody should be aware.*

The Greeks considered the earth the center of the universe. The Greeks before Aristotle described the motion of the heavenly bodies,

---

* You will find the story discussed in Arthur Koestler's, *The Sleepwalkers.* (Warning! Don't read Part Five, Chapter III—Koestler did not understand Newton.)

but it was Aristotle who codified all facts known to his time. His ideas were based on a distinction between heaven and earth. There was one set of rules for earth: On earth, everything had its proper place, the heavy objects below and the lighter media above. On earth, everything had its proper state, the state of rest. Forces may and do disturb this rest, but this is temporary and uninteresting.

With respect to the heavens, the set of laws were completely different. In heaven, the law was motion and, in particular, the most ideal motion: uniform motion on a circle. Almost all the heavenly bodies, the beautiful array of thousands of stars, are seen in this simple motion of perfect uniformity. That there are exceptions to the uniform circular motion is a sign of imperfection, but even these motions are approximated, and should and must be approximated by some combination of uniform circular motions. The sun, the moon, and the planets are the exceptions. These deviates move on circles whose centers move on circles; they move on "epicycles." But these approximations were not enough, they actually move on circles whose centers move on circles whose centers move on circles; at least as complicated as that. The Greek astronomers worked on this and could, with circles on circles on circles on circles, describe any motion of the heavenly bodies. By the first century after Christ, the system was complete and it was petrified by the astronomer Ptolemy in his famous book. A few hundred years later, Arab scholars called it "Almajest," the Majestic Work.

There were some who disagreed and, in particular, there was a remarkable man Aristarchus* of Samos. He lived in Alexandria around 200 B.C. and he asserted that the earth rotates on its axis and revolves around the sun, that the sun is at rest and that, if you make these assumptions, you greatly simplify the circles on circles on circles. Learned men of his time happened to disagree. He was ignored and his ingenious suggestion was lost. The only reference to this part of his work is to be found in Archimedes, who was kind enough

---

* It is a strange coincidence that his name means "the best beginning." The Latins say, "Nomen est omen."

to criticize him instead of ignoring him completely. (Nothing is as deadly as silence.)

I want to tell you a story which is not essential to the development of the Copernican system, but it is remarkable, a little ironic, and a little sad. The Greeks knew the distance to the moon reasonably accurately. This was done by the same method by which I can estimate your distance, by observing the displacement of your head on the background if I look at you first with one eye and then with the other. This method of measuring the distance to nearby objects is the parallax, and it won't work on distant objects unless your eyes are sufficiently far apart. By taking observations sufficiently far apart on the earth, the Greeks were barely able to notice a parallax of the moon and to get a reasonable estimate of its distance from the earth. The parallax of the sun was much too small to be noticed (besides, it is difficult to see stars near the sun).

Aristarchus made a suggestion which, in principle, is both ingenious and correct. He wanted to find the distance between the earth and the sun. If we see exactly one half of the moon, then the triangle defined by the sun, moon, and earth has a right angle at the moon (see Figure 1). Aristarchus measured as accurately as he could the time at which the moon was in its first quarter (we see it as 1/2) and the time it was in its last quarter (again we see it as 1/2). Then, by noting the difference in time it took the moon to travel from the first quarter to the last quarter, and the time it took to travel from the last quarter to the first quarter, Aristarchus calculated the angle α in the diagram. Using trigonometry he could then calculate the distance from the sun. Unfortunately, it was impossible for Aristarchus to determine the exact position of the moon when it was 1/2, so his result was in error by more than a factor of ten. This erroneous result the Greeks accepted. Even this inaccurate determination showed that the sun was much larger than the earth and he jumped to the conclusion that the tail should not be wagging the dog. But, as mentioned, the academy, the followers of Plato and Aristotle, disagreed with this unphilosophical wisdom.

Let me jump now to Copernicus. Copernicus was a traditional, rather plodding, and not too imaginative churchman. He got the job

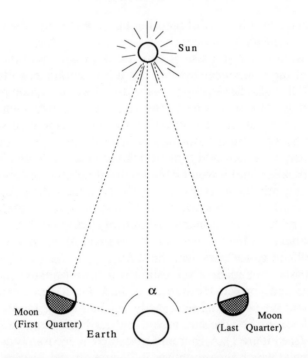

Figure 1.   Parallax gave the Greeks an estimate of the distance to the moon. Aristarchus observed the angle the moon makes as it travels from the first to last quarter and used trigonometry to calculate the distance from earth to the sun; he was only off by a factor of 10.

from the Pope to help in correcting the calendar and to eliminate the mistakes that had crept into the work of Ptolemy, which had been copied and recopied (not always accurately). As a result of reviewing Ptolemy, Copernicus adopted the hypothesis of Aristarchus. Copernicus was scared; he certainly did not want to do anything of which the church would not approve. The Bible clearly implied that the sun was moving. How else could Joshua have commanded it to stand still? Unable to reconcile such problems, Copernicus wrote down his understanding of the movement of the planets, but dared not publish his work. Finally his pupils got his ideas into print and

Copernicus saw it on his death bed. But even then Copernicus added an introduction, "Please don't take what I am writing seriously. It is nothing more than Mathematics and Mathematics is for the Mathematicians. It happens to be simpler to talk about the motion of heavenly bodies in this way, but I am not suggesting that you believe a word of it, and I am certainly not suggesting that it is real."* Even so, he was attacked, remarkably enough not by Rome, but by the Protestants who were very new, very revolutionary, and disliked any revolution different from their own. They believed the Bible verbatim and for them Copernicus was clearly anathema. Rome defended him. Of course, in the end the Copernican theory collided with Rome, but this collision is by no means the simple story of the tradition-bound church against the always infallible scientists.

The next actor in the drama was a Danish astronomer, Tycho Brahe, a man with a wild temper. He was given a little island, by the king of Denmark, on which he had an excellent astronomical establishment. Without telescopes, just by lining things up with great precision, he made measurements more accurate and more systematic than ever before. He took the Copernican theory seriously. He tried to test it by measuring the parallax of the brightest stars, which were presumably the nearest, with respect to the earth's orbit around the sun. He couldn't see a parallax due to the different position of the earth in summer and winter (or spring and fall). So he thought that either the earth does not move around the sun or the nearest star must be tremendously far away. He was, for an astronomer, not farsighted enough and he threw out the second possibility. Tycho Brahe could have been on the right track, had he been willing to assume that these little traveling objects were in reality as bright as the sun, and sometimes a thousand times brighter.

He believed he had disproved the proposition that the Earth moves around the sun. He modified Copernicus and said that the moon travels around the earth, the sun moves around the earth, but all other planets move around the sun, which itself travels in a circle.

---

* Given in a completely unverbatim translation.

In fact, the nearest star was further away than anyone was willing to guess: four light years. The distance from the sun to the earth is about eight light minutes, so it is not surprising that Tycho Brahe could not see the parallax. In 1830, Bessel did measure that parallax.

A young man who worked for Brahe was to have the most important part in the Copernican revolution. He was Johannes Kepler. His passion was to understand God's handiwork, and so he was interested in anything having to do with the planets. In particular, he was an astrologer; he wrote horoscopes for General Wallenstein in the Thirty Years War.*

You might think the practice of astrology is unscientific. To Kepler, astronomy and astrology were linked. In fact, Kepler's horoscopes have proved useful to science. Because of his notes, we know the exact position of the moon and the planets at a very specific time before an important battle. We can calculate that the earth's rotation is slowing down. This is actually due to the friction produced by the tides. Furthermore, Kepler's predictions often proved correct. Perhaps this was due to the circumstance that looking at the stars made him independent of current prejudices.

Kepler had theories, some of which were very nonsensical, and he wanted to test every one of them. He needed experimental data, so he got a job with Tycho Brahe who was then in Prague. Shortly afterward, Brahe died. Kepler then stole Brahe's papers from his heirs. Brahe's heirs wanted the honor of publishing these papers, but Kepler wanted to understand them. It is almost certain that Kepler was legally wrong. As a scientist, I think he was morally right.

In possession of Brahe's observations, Kepler was faced with a massive amount of data. So he did what any reasonable scientist would do and chose a particular problem. Furthermore, he chose the most difficult problem: the orbit of Mars, which deviated from a circle to the greatest extent. If he could explain the most difficult problem, then he could explain everything.† Kepler believed in Co-

---

* The Thirty Years War was from 1618 to 1648. There were plenty of questions an astrologer was supposed to answer and the *New York Times* was unavailable.

† This is an application of the old recommendation on how to catch a bird: Put salt on its tail.

pernicus; it appealed to his sense of simplicity. The earth rotates around its axis, the sun seems to move around the earth; Mars moves around the sun, so Mars appears to move on an epicycle of the third kind, a circle on a circle on a circle. But it doesn't. The earth doesn't move on a circle and Mars doesn't move on a circle. At least two more corrections were needed. Kepler wrote a treatise on the orbit of Mars. It is amusing because he starts many chapters by saying, "What a fool I've been when I wrote the last chapter. I must look at the orbit in a new way. . . ." The book is a paragon of scientific honesty and, therefore, is most illuminating concerning the scientific method. By contrast, the great Gauss corrected his errors in secrecy and is, therefore, practically ununderstandable.

After years of work, Kepler explained the orbit of Mars as a fifth-order epicycle. His calculations agreed within two times the uncertainty in Brahe's observations. He could have said that Brahe was wrong or he could have used a sixth-order epicycle to explain everything. But he didn't. At this point he threw out several years of work and started again. There are few scientists today who would have done the same thing. One of the greatest mysteries of this drama was Kepler's rebellion against the tradition which he followed with so much diligence.

I think the reason for this rebellion was Kepler's goal. He wanted to understand God's handiwork. The solution he found was not appealing. To use a sixth epicycle was too easy. It was not the only possible solution of its kind. Surely God had not put the universe together in such a haphazard way. He seems to have recognized the central principle of science: simplicity and the beauty of finding a unique and convincing answer.

Kepler decided to throw away circles and try ellipses instead. An ellipse is a circle viewed from the side.* It has two points inside,

---

* Consider all the circles at varying distances that will appear to you, looking with only one eye, to be of strictly the same size. They lie in a cone whose sides are traced out by rays reaching your eye from the circles' circumferences. If you cut this cone by a plane parallel to the circles you get, of course, a circle. If the plane is oblique, you get an ellipse. If you turn the plane so that it becomes parallel to one of the rays, the line of intersection is a parabola. This would be the orbit of a planet that approaches from infinity and

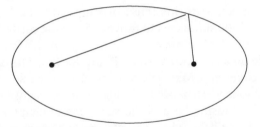

Figure 2.   For any point on the perimeter of an ellipse, the sum of the distances to the two foci is a constant. If the foci come together at one point, the ellipse becomes a circle.

called foci. For any point on the ellipse, the sum of the distances from that point to the foci is a constant (see Figure 2). Once Kepler had ellipses, the rest became easy.

In the end Kepler arrived at three laws about the motion of planets. The first law states that planets revolve around the sun on ellipses with the sun at one of the foci of the ellipse.

To understand the second law we take some time interval, say, one week. Then consider the figure formed by the line between the sun and the position of the planet at the beginning of the week, the line between the sun and the position of the planet at the end of the week, and the path that the planet has traveled. This figure will have an area which will be constant, the same area week after week (see Figure 3). The law, then, is that a planet sweeps out equal areas in equal periods of time. This determines the varying speeds with which the planet travels: slow, when far from the sun (aphelion), fast, when close to the sun (perihelion).

The third law discusses a constant. If D is the big diameter of the ellipse for some planet and P is the time it takes that planet to

---

recedes to infinity, having lost all its velocity. If you turn the plane even further, you get a hyperbola. This would be the orbit of a planet that comes from and returns to infinity, but retains some velocity at all times. The Greeks knew of all these "conic sections" so Kepler could find them in the literature.

Planetary orbit

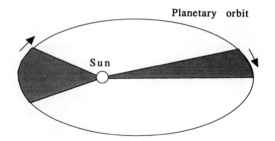

Figure 3. A planet does not move at constant speed about the sun; it moves fastest when nearest the sun. Kepler correctly noted that its velocity changes so that the area swept out by the planet (shaded areas) in any given time interval is a constant, no matter where the planet is in its orbit.

make a complete revolution (or period) around the sun, then D times D times D divided by P times P (or D^3/P^2, as the mathematicians would write it) is a constant. It has the same value for each planet that revolves around the sun. That means that if you calculate D^3/P^2 for D equal to the diameter of the earth's orbit about the sun and for P equal to the period of the earth about the sun (one year), and then compare that to D^3/P^2 with D for Mars's orbit about the sun and P for Mars's period of revolution about the sun, they are the same.

Kepler assembled these results in a book. He wrote an introduction which was a strange mixture of modesty and incredible pride. It said something to the effect: "I believe nobody will read this book for 100 years, but I don't care. God had to wait 6,000 years for somebody to understand his work." It was just about a century later that Newton read Kepler. He understood the laws even better than the author.

Kepler was a mystic. He was a complicated person. He had a fixed view of his goals and worked incredibly hard to reach these goals. His laws had a profound effect on the development of science.

Galileo was simpler. He was a propagandist and, although he was a good scientist, he was not dedicated to one goal as Kepler was. Galileo was the first to confront the question: If the earth rotates and moves in its orbit, why does it appear to us as standing still? His interest in physics was first aroused when he saw a lamp swinging in church. He noticed that the period of the swing was independent of the amplitude, that no matter how high the lamp was swung, it took the same amount of time for the lamp to complete a swing. Galileo studied the velocities of falling objects as a function of time. The story goes that he dropped objects from the Leaning Tower of Pisa to confirm his conclusions: heavy and light objects fall in the same time. It is one of the unfortunate omissions of history that this experiment did not actually happen.

Galileo was a Copernican, but he kept quiet about it until a Hollander invented the telescope. Galileo then reproduced this new invention and probably his telescope was the best of his time. Then he did the obvious and turned his telescope toward the sky. He published a book,* *A Messenger from the Stars.* Galileo wanted his book to be widely read, so he wrote his book in Italian instead of scholarly Latin. It came complete with pictures and it was sensational. Galileo ignored Copernicus's warning that his theory was only mathematics; Galileo followed Copernicus's formulae, not his words.

In his book Galileo described what he saw through his telescope. He described the craters and mountains on the moon. He described the phases of Venus, where Venus sometimes appears as a full disk and sometimes appears as a sickle. He described the four bigger moons of Jupiter, a model of the Copernican planetary system. Galileo looked at the Milky Way and stars too faint to be seen with the unaided eye. Here was another bone of contention. According to some churchmen, the stars were ornaments. Why should God have created ornaments that could not be seen?†

The Jesuits and also one of the cardinals, Maffeo Barberini,

---

* The actual title is *Sidereus Nuncias.* The usual translation is *Starry Messenger.* This translation is not much better than mine and it is stilted. I believe Galileo would have used the title given in the text had he written in English.
† WT: The answer is easy. He liked astronomers.

were interested in astronomy. In fact, Barberini wrote a poem in honor of Galileo. The Pope was not pleased, however, and he was not completely to blame. The church was a big organization, perhaps even including red tape. It had the duty to explain everything. If Copernicus was right, everything would have to be reexplained. If it turned out after all that Tycho Brahe was right and Copernicus was wrong, then the church would again have to take everything back. This was simply too much to expect from a big organization. The church did not reject Copernicus out of hand. Galileo was informed that, if he wanted to discuss the Copernican theory, he could do so provided he presented both sides of the argument.* The church acted like a good conservative scientific society. Certainly Galileo was better treated than Aristarchus. He was permitted to discuss his theory and he was not ignored. Galileo obeyed the dictates of the church for the time being.

Barberini, the cardinal friend of Galileo, became Pope Urban VIII. Galileo went to his friend the new Pope, claiming he had a proof of the Copernican theory. His proof was that the rotation and the forward motion of the earth caused the tides. The water could simply not keep up with the earth's complicated motion. Therefore, Galileo claimed, "I have explained the tides on the basis of the Copernican theory and so the Copernican theory must be correct."

Unfortunately Galileo made two big errors in his explanation. First, the theory that Galileo proposed would explain only one high tide each day and there are two high tides. Kepler, after years of work, refused to add another epicycle to correct a very small deviation between calculation and observation. Galileo was willing to ignore the difference between one and two! (In the Mediterranean, the tides are hard to notice.)

---

* This was repeated centuries later by the instructions of Clark Kerr, then President of the University of California, to the University: "Discuss any political theory, but present both sides." The radicals rebelled not in the name of truth, but in the name of "free speech;" only their side was to be heard.

The second error is even more remarkable than the first, in a way, because Galileo was willing to ignore a principle that he himself discovered and that bears his name. Galileo's principle is that uniform forward motion feels the same as being at rest. If you are traveling on a ship and you do not look out the window, except for the occasional rocking of the ship, it feels as if you were at rest. A hammer dropped from the mast of a moving ship lands at the foot of the mast—even while an observer on the shore says it traveled on a parabola. This principle is very important, if one is to believe the Copernican theory. If the earth is moving forward as Copernicus says, then according to Galileo we aren't left behind. Indeed, the forward motion should not be observable and should not influence the tides. But Galileo ignored his own theory when it suited him.

Galileo planned to publish his proof in a book which he intended to call the *Flux and Reflux of the Tides.* Had he used this title, with its obvious emphasis on a fallacious proof, his reputation would have suffered. The Pope, who proved in this situation to be a better scientist that Galileo, convinced him not to use this title. Instead Galileo wrote a book about the Copernican and Aristotelian systems of the universe, called the *Dialogue on the Two Great Systems of the World.* It is a conversation between three people, Salviati, who had all the answers, Sagredo, who asked all the reasonable questions, and Simplicio. Simplicio, as his name indicates, asked all the silly questions. The others, with great patience, explained everything to Simplicio and then added, "God can do anything, so maybe you are right."

It happened, hardly by accident, that all the Pope's arguments came from the mouth of Simplicio. This was too much for the Pope. Galileo was called to the Pope. At first he did not come, claiming ill health. It was true that he was old and not in the best of health, but finally he went. He was intimidated, but certainly not treated roughly. He was made to abjure the Copernican theory and was forbidden ever to discuss it again. He was put under house arrest. Galileo got bored and so he wrote down his experimental results. He is certainly indebted to the Pope. First for making him a martyr, second for forcing him to publish his truly scientific results. Without these, Galileo's name would not be so well remembered today.*

Science is a matter of taste. As far as my taste goes, Kepler was a better scientist than Galileo. Galileo did experiments that others would or could have done. Kepler did things other people would not do. He believed that the motion of the planets contained the secret of creation and he was determined to discover that secret. In addition to his three laws, he also suggested ideas which were worthless but of which he was equally proud. This does not detract from him as a scientist. He was willing to throw away years of work so that he could find solutions which were simple and elegant. Furthermore, without his laws, it is doubtful that Newton could proceed with his work.

At any rate, Copernicus, Kepler, Galileo, and Pope Urban VIII created the background for the scientific revolution that would follow in the next few decades.

QUESTIONS

1. Galileo reasoned correctly that, because Venus had phases like the moon, its orbit must be closer to the sun than the earth's orbit. What was Galileo's reasoning?

2. The group of stars closest to us is called Alpha Centauri. The brightest star in that group appears to be 70 billion times less bright than the sun. We now know that it is only 20% less bright than the sun. How do the distances to the sun and to Apha Centauri compare? (The sun is 500 light seconds or approximately eight light minutes away; in similar terms, how far away is Apha Centauri?)

---

* ET could use a reasonable sentence of house arrest. Of course, today it would not be so effective because of the invention of Bell. ET suffers from severe telephonitis.

3. Astronomers measure distances, not in light years, but in parsecs—the distance at which the parallax to the object is one second of arc. What is the distance of Alpha Centauri in parsecs?

4. Estimate the ratio of brightness of the new moon to that of the full moon.

5. Galileo tried to measure the speed of light between mountain tops. He failed. Could he succeed today?

Chapter 4

NEWTON

*In which the reader will find out about the law
of motion which applies equally
on earth and in heaven.*

According to the Greeks, rest was the natural state on earth.* This
had two advantages: it appeared obvious and it agreed with the notion
that the earth itself was at rest in the center of the universe. But
Galileo believed in the Copernican theory. Although he did not state
it, he implied that either rest or uniform motion on a straight line
was the natural state on earth. Certainly uniform motion on a circle
could not occur without a force inward. Galileo, however, still failed
to contradict the notion that uniform circular motion was natural
in heaven and did not require a force pointing inward. A modern

---

* This contrasted with the motion of the stars and the planets; movement was "natural"
  for heavenly objects.

student may find it "logical" to postulate that uniform motion is natural everywhere. He may argue that since Galileo has stressed the similarity between landscapes on the moon and earth, he should have taken the next step to extend this similarity to the laws that apply. The 20/20 vision of hindsight sees everything except, in many cases, the established conviction that had prevented such vision in the past.

You can feel the force connected with circular motion if you whirl an object on a string. It is here that Newton began his discussion. What was most important was that he considered the minute details of this motion.

Suppose that **r** is the position of the planet moving (for the sake of a simple example) on a circle. Now what is its velocity? The velocity is the change of position divided by a change in time. I write this as $\Delta \mathbf{r}/\Delta t$. If I take a long period of time and divide that into the distance that the planet has traveled in that time, I get an average velocity. It may be that at some time the planet has traveled much faster than the average velocity I have calculated, and at other times much slower. Now what Newton (and, by the way, also Leibnitz) did was to let Δt get small. He "took the limit of $\Delta \mathbf{r}/\Delta t$" as Δt got very small. This limit we write as $d\mathbf{r}/dt = \mathbf{v}$, where **v** is the velocity at one moment. This idea of taking the limit of $\Delta \mathbf{r}/\Delta t$ is the idea of differentiation.

Newton's starting point was: What really matters is not the velocity, rather the change of velocity, which is called acceleration. When the proud new owner of a car says that it has good acceleration, he means that it picks up speed quickly. (It is even more important to have brakes which can produce negative acceleration in an emergency. In fact, physicists are not one sided; by acceleration they mean either that rate at which speed is increased or the rate at which speed is decreased—we call this positive and negative acceleration.) Acceleration, to a physicist, means the change of the velocity in a certain period of time. It is permissible to play the same game with acceleration that we played with velocity. Acceleration is $\Delta \mathbf{v}/\Delta t$, but to be accurate we want Δt to be very small, so we take the limit and write **a,** the acceleration to be equal to $d\mathbf{v}/dt$.

I said that **v**, the velocity, is equal to $d\mathbf{r}/dt$, so acceleration is $d(d\mathbf{r}/dt)/dt$ or, as it is usually written, $d^2\mathbf{r}/dt^2$; the "2's" are to remind you that we have differentiated twice (although the placement of the twos is a mere convention). You might say at this point that I am as bad as the Greeks with their epicycles, but at least I will not go further and differentiate a third time at this stage in the discussion (nor in the following chapters).

It was the idea of differentiation that Newton was interested in as a student in Cambridge (England).* At that time a disaster occurred, caused by pollution (beyond the imagination of a modern ecologist). Unsanitary conditions allowed plagues to occur and during one of these dreadful epidemics, Cambridge was closed. Newton returned to his native village, Woolsthorpe. It is there that Newton, when he was 21, made his great discoveries. (We have no proof that he did not do it under an apple tree.) He was interested in forces; he guessed that force would be proportional to acceleration.

I have cheated a little when I described velocity and acceleration. You will notice that I nonchalantly wrote both velocity and acceleration as boldface, by which I implied that they are both vectors. This is not surprising. The velocity is the difference of two positions, which are vectors, divided by a number, time.† Therefore, velocity is a vector. Similarly, acceleration is the difference of two vectors, velocities, divided by a number, time. So acceleration is also a vector.

The idea that force and acceleration are proportional is a natural idea. Both are vectors. A force causes change from uniform motion. Acceleration describes the change from uniform motion. Newton asserted that $\mathbf{F} = m\mathbf{a}$, where m is the mass. The meaning of this equation is that two objects have the same mass if the same force produces the same acceleration. Similarly, one object has half the mass of another if half the force produces the same acceleration.

---

* Cambridge (Massachusetts) already existed, but did not compete.
† We have seen that time is a component of a vector in four-dimensional space-time. We have also seen that, if we deal with small velocities (and that is what Newton did), time is practically an invariant. So one can describe it by a simple number.

Actually, this is most plausible if one compares an iron bar with half of the same iron bar. It is a little less obvious if one compares an iron bar with a wooden stick.

Using the idea of differentiation and the law $\mathbf{F} = m\mathbf{a}$, we can discuss uniform motion on a circle. Suppose a planet moves a certain distance Δr along the circle. Then we can describe this distance as a fraction of the distance around the circle. The angle $\Delta\alpha$ in Figure 1 equals Δr divided by the radius of the circle. This means that we don't describe $\Delta\alpha$ in degrees, but in "radians." If we traveled one complete circle, $\Delta\alpha$ would equal circumference/radius = 2π.

Now we would like to know the rate of change of α. We say $\omega = \Delta\alpha/\Delta t$ and we call ω, reasonably enough, the angular velocity, the velocity at which the angle is changing.

We have chosen ω to suit our discussion. In a time Δt, the distance a planet has moved on a circle will be $\omega R \, \Delta t$, where R is

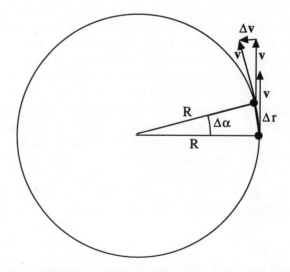

Figure 1. For uniform circular motion, we can relate the velocity **v** to the angular velocity and obtain the inward acceleration, v^2/R, needed to keep the object moving in a circle.

the radius of the circle and we have chosen a short time Δt to justify the rules of differentiation. The size of any two velocities will be the same, because the planet is moving at uniform speed. The direction of the velocity will change and it will change just as rapidly as the direction of the line between the center of the circle and the position of the planet is changing. Therefore, the triangle formed by the two radii and Δr is similar to the triangle formed by the two velocities, v and Δv. So we get $\Delta v/v = \Delta r/R$. But $\Delta r = \omega R\, \Delta t$, so that $\Delta v/v = \omega$ Δt and $\Delta v/\Delta t = \omega v$. Now since $\Delta r = \omega R\, \Delta t$, we have $\Delta r/\Delta t = \omega R$ or v $= \omega R$. So we get for the acceleration $\mathbf{a} = \Delta v/\Delta t = \omega v = \omega \omega R$ $= \omega^2 R$. The acceleration is $\omega^2 R$ (or v^2/R) and since $\mathbf{F} = m\mathbf{a}$, the force is $m\omega^2 R$.*

In order to distill the exact laws of planetary motions from observations, Kepler started with the orbit of Mars, which differs most strongly from a circle. In order to understand these laws, it is best to start from the opposite end: the uniform circular motion of planets, which is the simplest case, and is not so far different from the actual behavior of planets. Kepler's third law can be stated for this case by saying that R^3/T^2 is the same for all planets. As in our earlier discussion, R is the distance between the centers of a planet and the sun. I want to define T as the time in which the planet covers an arc whose length is R.† Since the distance covered in a time t is $\omega R t$, we have $\omega R T = R$ or $\omega T = 1$.

From Kepler's third law, which compares the motions of planets at different distances, it should be possible to find out how the force that keeps the planets on their orbits changes with their distance

---

* The clever reader will have noted that we must have been taking small increments to allow us to use Δt for dt, and Δv for dv, and that we have avoided the use of sines and cosines in our calculations and arguments by using vectors which are at right angles to each other.

† I almost would like to call this primitive formulation (equivalent to Kepler's famous early discoveries) Kepler's first law. But some things, like the alphabet and Euclid's fifth postulate, are unchangeable parts of our heritage. It may have been the first law which Newton understood in Woolsthorpe. (Others arrived at the same understanding before Newton's great work was published.)

from the sun. Instead of Kepler's law (R^3/T^2 = constant), we can say that $R^3\omega^2$ is the same for all planets. We will write for this constant $K = R^3\omega^2$. The force needed to keep a planet traveling with uniform velocity in a circle is $F = m\omega^2 R$ or, using the constant, $F = mK/R^2$. This means that the attraction to the sun must be proportional to the mass of the planet and to $1/R^2$.

Galileo saw that Jupiter had four moons. Does the same law hold for the moons of Jupiter? The answer is both yes and no. It is true that $R^3\omega^2$ is a constant for the moons of Jupiter, but it is not the same as the constant for the planets. The constant depends on the central body. Thus K is different for the motion around the sun, different for the motion around Jupiter, but it does not vary for the moons of Jupiter and it does not vary for the planets going around the sun.

It's a pity that the earth has only one moon. If it had another moon, we could test the $1/r^2$ law. Here the apple fell on Newton's head, at least figuratively. Galileo already measured the acceleration of the apple due to the earth's gravity. Newton, using the formulae we have derived, could calculate the velocity needed to keep the apple on a circle going around the earth. Then he compared the apple (turned satellite) and the moon and he found that the $1/r^2$ law held. Of course, in the space age we do not need apples, we use (slightly more expensive) satellites which move, without friction, outside our atmosphere and encircle our globe in as short a time as 1.5 hours.

The statements of Newton are rounded out and made consistent by adding that the gravitational force is proportional to the mass of the attracted as well as the attracting bodies. The general formula is $F = G(m_1 m_2/r^2)$, where G is the universal gravitational constant* and the two interacting masses are m_1 and m_2.

---

* Remarkably enough, G was measured in 1798 by Cavendish to four figures as 6.754 $\times$ 10^{-8} cm^3/g sec^2 and not until relatively recently was the accuracy improved to 6.673 $\times$ 10^{-8} cm^3/g sec^2. This was because we have no independent information on the masses of the earth or of the celestial bodies.

That two times the quantity of matter should attract (and be attracted) twice as strongly is again plausible. It is much more surprising that "quantity of matter" can be compared in a similar way, never mind whether we consider acceleration (as in the formula $F = ma$) or gravitation [as in $F = G(m_1m_2/r^2)$]. If the mass is doubled, so is the accelerating force of gravity. If a quantity of wood is as hard to accelerate as a quantity of iron, that quantity of wood will also exert the same gravitational attraction as the quantity of iron. This identity of the gravitating and the inert mass has become the starting point of Einstein's explanation of gravitation, of which we shall speak later.

What has been said means that the law of gravity holds not only for the sun and Jupiter and the earth, it holds for everybody and everything. I am attracting you, you are attracting me. I can even measure the $1/r^2$ law using a simple device. From the ceiling of a garage, or any empty room, suspend a pole with equal weights hanging from either end of the pole as in the diagram. Everything should be balanced. Then if I move a large object (e.g., William "the Refrigerator" Perry*) near one of the weights, the whole contraption will rotate (see Figure 2). The large object has a force of gravitation and even though this force is small (since my large object was not nearly as large as the earth), we can see the weights rotate because they are sensitive to small forces.

The change introduced by Newton was thorough. It is no longer true that on earth the law is to be at rest, in heaven to move in a circle. $F = ma$ is true in each case. Even the cause of motion or its change is the same. For instance, the $1/r^2$ law holds for planets and for ants. The great philosophic change that was introduced is the idea of "One World" or, I should say, "One Universe." The most remarkable thing about the universe is not that it is so big but that it is so uniform.

---

* ET: What "Refrigerator"?
  WT: Clearly ET does not live in Chicago nor does he spend enough time watching NFL football. Pity.

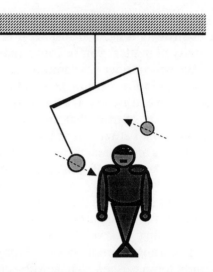

Figure 2.   If a pole is suspended from the ceiling with equal weights hanging from its ends, a large object near one of the hanging weights will exert enough gravitational force to cause the pole to rotate toward the object.

The expression $F = G(m_1 m_2/r^2)$ is symmetric. I attract the earth as much as the earth attracts me. Of course, the effect on the earth is not as noticeable, because the earth has such a large inertia compared to mine.

I attract you as much as you attract me. Newton generalized this idea. He said, "Any two bodies which interact undergo equal but opposite forces, and these forces act along the line between the two bodies." I am assuming that these bodies act only on each other. The consequence of this assumption is a law "conservation of momentum" (which we shall define presently). When any quantity is conserved, that is, unchanging, we obviously have a simple law. It is the purpose of science to find simple laws.

I suppose that the force on the first body is $\mathbf{F}_1 = m_1\, d\mathbf{v}_1/dt$, where m_1 is the mass of the first body and $d\mathbf{v}_1/dt$ is the acceleration of the first body. Similarly $\mathbf{F}_2 = m_2\, d\mathbf{v}_2/dt$ is the force on the second body. Then Newton's "action–reaction" law simply says $F_1 + F_2$ = 0 or $0 = m_1\, dv_1/dt + m_2\, dv_2/dt$.

Now I will do something crazy. Instead of writing $m_1\, dv_1/dt$, I will write $d(m_1v_1)/dt$. This isn't really so terrible, because differentiation is the rate of change and m_1 doesn't change. It will not make any difference, then, whether I multiply m_1 by dv_1/dt or whether I first multiply m_1 by v_1 and then go through the limiting process. So I can rewrite my equation as $d(m_1v_1)/dt + d(m_2v_2)/dt = 0$.

I will change my equation once more. This time I will add the quantities m_1v_1 and m_2v_2 and then go through my limiting process. So I now have $d(m_1v_1 + m_2v_2)/dt = 0$. The quantity m_1v_1 is called momentum or the quantity of motion. It is how much, m, times how fast, v, and is usually denoted by the letter p (for some unknown reason).

What my last equation states is that the total momentum $m_1v_1 + m_2v_2$ does not change with time provided only m_1 and m_2 interact. It also means that if v_1 changes, then v_2 must also change; and v_2 must change in such a way that $m_1v_1 + m_2v_2$ stays the same. This is the law of conservation of momentum. The law actually holds quite generally. It holds for any number of bodies. If I have 100 particles interacting, then the momentum of a particular particle does change. But the sum of the momentum values of the 100 particles remains the same. This law follows from an easy generalization of the statements made. It has been verified with high accuracy.

For the Greeks, the natural state of things on earth was rest. For Newton the natural state was rest or uniform motion. Then why are all of us not moving all the time with high velocities? The answer is connected with another conservation law, the conservation of energy. Earlier I discussed potential energy. A ball of mass m at a height h has a potential energy hgm, where g is the gravitational acceleration at the position of the ball. Now suppose I drop the ball and look at it just before it hits the ground. Its potential energy is gone, so what happened to its energy? The ball doesn't have potential energy, but it has kinetic energy, the energy of motion. The ball is traveling at some velocity v, and if the ball, by a perfectly elastic collision, reverses its velocity, it can ascend back to the original altitude h. The same is seen in case of a swinging pendulum. Potential energy and kinetic energy continue to change into each other during each swing.

Actually, the kinetic energy (as we shall verify below) is $mv^2/2$. If a ball was dropped from a height h, then the conservation law demands $mv^2/2 = mgh$ or $v^2/2 = gh$.

Let us assume the statement to be true* for h. It is also true for the slightly greater distance of descent, $(h + dh)$. Then the difference in potential energy is dhg. In the first case we have $v^2/2$ and in the second case we have $(v + dv)^2/2$, because in the added small distance, dh, the ball has gained a small additional velocity dv. Now we want the difference

$$\frac{(v + dv)^2}{2} - \frac{v^2}{2} = \frac{v^2}{2} + \frac{2v\,dv}{2} + \frac{(dv)^2}{2} - \frac{v^2}{2}$$

The two terms $v^2/2$ cancel and the term $(dv)^2/2$ is neglected, because dv is small and $(dv)^2$, which is something small times something small, is something terribly small. We are left with $2v\,dv/2$. Now my assumptions remain consistent if and only if $d(hg) = v\,dv$. I will divide everything by dt to get $d(hg)/dt = v\,dv/dt$. But, $d(hg)/dt$ is equal to $g\,dh/dt$, since g is a constant. Because dh/dt is the velocity v, on the left-hand side we get gv. On the other side, dv/dt is the acceleration a, which in this case is due to gravity and so is equal to g. So our last equation is reduced to the identity $vg = vg$.

We began this for a drop through Δh, and the result, $gh = v^2/2$, is obvious. For each additional Δh, the result remains valid. So we can accept $gh = v^2/2$ for all values of h.

You may wonder what happens to the energy when the ball hits the ground and the rebound is not elastic. If I do not drop a ball,

---

* WT: There is no use in objecting to this apparently unscrupulous way of proving a proposition. From sad experience I know that ET invariably jumps to conclusions and then looks where he has landed. Do you prefer a persistent snail to an occasionally remorseful grasshopper?

but a piece of chalk, then the energy is dissipated into breakage and heat. We will discuss this in Chapter 6. For the time being, we have the answer to our question: Why are we not moving constantly? We have a certain amount of mechanical energy. This amount is conserved. Additional energy is not generated freely. Furthermore, things such as friction use up this energy.

It was actually Kepler who noticed the first conservation law, only he called it by another name. We can now call it Kepler's second law or the law of the conservation of angular momentum.

The second law says that the line between the sun and the planet sweeps out equal areas in equal periods of time. For uniform motion on a circle, this is obvious. For ellipses the statement is not obvious. Assume that the planet moves with constant velocity on a straight line, which would be the case if the sun exerted no force on it. Then in equal times, we are considering the areas of two triangles whose bases are equal and, since the bases lie on the same straight line, the altitudes are also equal. The areas of the two triangles must then also be equal.

But the planet does not travel on a straight line. I have drawn in Figure 3 the path when the planet travels from A to B to D instead of from A to B to C. The change in velocity is directed toward the sun, as indicated by the heavy arrows. Now you will note that the triangle SBC has the same base as the triangle SBD, namely the line SB. The two triangles also have the same altitude, since DC is parallel to SB. So the two triangles must have the same area. But the triangle SCB is equal in area to both triangles SDB and SBA. So these must also be equal in area and I have proved Kepler's second law. [Actually, I have cheated, because the line CD should be pointing toward the sun instead of being parallel to SB. In small enough time intervals, this error is very small, on the same order as $(dv)^2$, and we can neglect it.]

In the derivation, the strength of solar attraction, or the dependence of the force on the distance, did not enter. We only used the circumstance that the force is directed toward the sun. That being so, the sun does not exert a torque on the planet: The torque is the

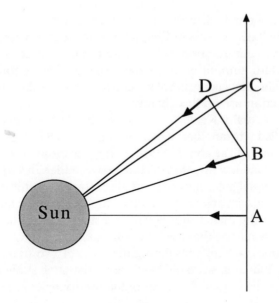

Figure 3.   So long as the force on a planet is directed toward the sun, Kepler's law of equal areal velocities must be correct.

vector product of the vector representing the force and the vector representing the displacement of the planet from the sun. Since their direction is the same (and the included angle must therefore be zero), the torque vanishes.

The angular momentum is defined as the vector product of the momentum and the displacement from a fixed point, in this case the sun (which is almost a point and almost fixed). One can then show that the time rate of change of the angular momentum equals the torque, and for a force directed toward a fixed center, the conservation of angular momentum and Kepler's second law follow easily.

It is just a little more difficult to generalize the conservation of angular momentum to any number of particles interacting with forces along the lines connecting each pair ("central forces") and not subject to outside forces. Conservation of angular momentum played an important role in the discussions of the formation of the solar system.

Of course, we have proved Kepler's second law only for thin slivers of triangles, slivers which still have to be added up or "integrated." The art of integration consists of adding many very small quantities. Then one increases the number of terms and decreases the size of each term. In this way one can calculate areas, for instance, those which occur in Kepler's second law.

The use of limiting processes in differentiation and integration is similar. Both were independently developed by Newton in England and by Leibnitz in Germany. Furthermore, integration and differentiation are "inverse operations," that is, integration undoes differentiation and vice versa, just like multiplication by, say, 3 undoes division by 3.

Newton originally assumed in his calculations that the sun and the planets had all their mass concentrated in a point at their center. Of course, in reality, the earth's mass is spread out in a sphere (well, not exactly a sphere). Newton wanted to prove that the assumption that all mass was located in the center of the body would not upset the work he had done.* In the end Newton did prove this, but only with complicated calculations. We now have a simpler way around this difficulty, a way which was not invented for many years after Newton's death. The simple way is to use Faraday's lines of force.

Faraday† assumed that the lines must start at infinity and can only end in the mass point. Furthermore, the number of lines are proportional to the mass of the attracting body. Newton's problem becomes simple. The planets are spheres (almost) and are symmetric, the lines of force are symmetric around the center. Whether the mass

---

* The many years that passed between the early conception in Woolsthorpe and publication by the famous professor in Cambridge may have been partly due to such worries.

† He really used these ideas, not in connection with gravitation, but in analogous problems in electricity and magnetism.

is considered as a point or the mass is distributed in spherically symmetric shells makes no difference.

Suppose that we have a center of attraction. Then we can visualize a force obeying the $1/r^2$ law, due to the attraction of the center, by drawing lines emanating from the center. The lines will be of uniform angular density, because the force is the same in any direction. To find the strength of the force at any position, we simply look at the number of lines crossing a unit area. For example, if in Figure 4 we have a sphere of radius 1 and sphere of radius 2, both drawn around the center of attraction, then the number of lines has stayed the same, but the areas ($4\pi r^2$ is the area of a sphere) have

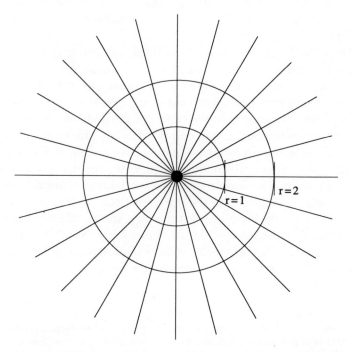

Figure 4.   The circles represent spheres. The area of the inner sphere is 4π and that of the outer is 16π. However, the same number of lines emanating from the center of attraction must cross both spheres.

increased from 4π to 16π. Thus, the force is proportional to $1/4\pi$ and to $1/16\pi$ and we see that the $1/r^2$ law holds.

Of all of Kepler's laws the first was hardest to find and is hardest to prove. Even Newton had some difficulties. There is a story on this subject and, in contrast to most stories about Newton, this one is true.

Comets, which on rare occasions appear as spectacular visitors among the stars, seemed to come from a region beyond our knowledge and return into the unknown distance. In Newton's time, it had been assumed that they move on parabolas—ellipses with an infinite major axis. A British astronomer, Halley, noticed indications of periodicity. Perhaps the comets were moving on ellipses with a long but not infinite axis. According to Kepler's third law, the period should, indeed, be long.

So Halley went to the best authority, Newton, and asked (paraphrasing), "It is said that the elliptic orbits follow from the inverse square law of gravitation. Can you prove it?" Newton said, "Yes, I proved it in Woolsthorpe." He said he would have the proof by the next day. The next day came without a proof. Newton couldn't reproduce it. After two weeks he finally reconstructed the proof and gave it to Halley.

The story has two sequels: one which took more time, the other less. The longer story is that Halley did predict the return of an important comet. The comet appeared on schedule after the death of Halley. It is now appropriately called "Halley's Comet."

The shorter story is more important. It is said that the experience with Halley made up Newton's mind. Now he must write down everything. This was the decision that led to the publication of the most important book on science since Euclid. The title is *Philosophiae Naturalis Principia Mathematica,* or, briefly, *Principia* (the mathematical principles of natural philosophy). It revolutionized science, mostly (but not always) for the better.

It proved that there is *one* law in the universe.

It established a firm link between physics and mathematics.

It eliminated thoughts about action spreading through space and substituted action at a distance.

It erected a firm framework of space and time.

It suggested that particles should be considered as the explanation of physical phenomena.

We still believe the first two of these propositions. The last three have been modified with a delay of one or two centuries.

Principia turned poor Newton into a superscientist. His great work was done, but his drive was not satisfied. He became master of the Mint. He worked on alchemy. He wrote theological papers. (I actually heard of a young man who knew of Newton not as a physicist, but as a theologian.)

Would physics be different today without Newton? I guess the answer is *NO*. What he accomplished was done with brilliance. But the ideas he put forward in a finished and satisfying form had been actually in the air.

In my opinion, the man who really made the difference was Kepler. Without his ellipses, human thought about the heavens may not have been clarified for many more years.

Newton seems to have understood his role better than some of his contemporaries. He described himself as a child who has found some pretty shells on the shores of the great unexplained Ocean of Knowledge. He left the shells more beautiful than they were when they first came to his notice.

QUESTIONS

1. The Archimedes–Galileo question: Given two spheres of equal radius and weight, one of solid aluminum and one hollow gold, can you find out which is which.

2. I consider an apple travelling around the Earth, just as Newton did. How will the time it takes the apple to make one revolution depend on its distance from the earth? In particular, how far up must the apple be in order to make one revolution in 24 hours? (In this case, if the apple is above the equator, it would appear to be above

the same spot on the earth at all times. Satellites, which do exactly this, are used to transmit radio and television signals. Such a satellite is called a geosynchronous satellite.)

As an interesting aside, note that if the actual mass of the earth were distributed uniformly out to this radius and the earth were to continue to rotate once in 24 hours, the earth's gravitational field would barely suffice to hold the surface against the centrifugal force and the tides would become disastrous.

Chapter 5

"HYPOTHESES NON FINGO"

In which the reader may learn how Newton's simple facts on gravitation were superseded by the imagination of Einstein.

The title of this chapter is a boast of Newton's: "I do not make hypotheses." More than two centuries after the publication of the *Principia,* Einstein shook this unshakable edifice. There is little question that he did make some hypotheses. They turned out to be incredibly fruitful.

Remember that I said that $(ct)^2 - r^2$ is an invariant? This statement is the modern basis of the geometry of space and time. It contradicts one of Newton's "facts" according to which time proceeds everywhere in the same manner, unaffected by space, not coupled to space.*

---

* A fact is a simple statement that everyone believes. It is innocent, unless found guilty. A hypothesis is a novel suggestion that no one wants to believe. It is guilty, until found effective.

We can play the same game with an invariant introduced by Einstein which collects into a single law three laws of conservation: conservation of mass; conservation of energy; and conservation of momentum. The new invariant is $E^2 - (cp)^2$. E is the energy of an isolated system and p its momentum, that is, its velocity times its mass (c is the velocity of light). I am going to tell you what this invariant is.

$E^2 - (cp)^2 = (m_0 c^2)^2$, where m_0 is the mass of the system. I will explain the subscript "0" in m_0 shortly.

We want to know that all the quantities in the equation are of the same kind, that is, they are measured in the same units. I said that the kinetic energy of a particle is $(mv^2)/2$. So energy must have "dimensions" of (mass) $\times$ (velocity) $\times$ (velocity). But c is a velocity and p has the "dimensions" (mass) $\times$ (velocity), so cp is the same kind of quantity as E. Of course, $m_0 c^2$ also is the same kind of quantity.

For our first invariant, $(ct)^2 - r^2$, we found that r was different for different observers. This was not so surprising. What was surprising was that time was also different for different observers. This difference in time was only perceptible when the velocities were very large. For our new invariant, we have an analogous situation. Instead of r, we have p; r is a vector with three components, p is a vector with three components p_x, p_y, and p_z, or mv_x, mv_y, and mv_z which are the components of the vector. Instead of time, we have energy. Now every event can be characterized by four numbers, x, y, z, and t, that is, every event is a four-dimensional vector. Of course, with a change of observers, the x, y, and z part of the event vector will be confused with the t part. The same holds for E and p, which are, again, components of a four-dimensional vector. How much of this vector is E and how much of it is p depends on the observer. For example, suppose I were in an airplane and had my pen with me. Then for me the momentum of the pen is zero. If I happen to drop my pen out the window, then to an observer, say, a man parachuting (and having the same velocity as the air), the pen will have a very great momentum. The parachutist had better dodge the pen, otherwise he might be hurt. So it is clear that momentum is different for different observers.

What about E? Does it change for different observers? Suppose the momentum is zero, that is, we are in a coordinate system where the system is at rest, $v = 0$. Then our equation becomes $E^2 = (m_0c^2)^2$ or $E = m_0c^2$. This is a well-advertised equation. You can now see that m_0 stands for the mass at rest (or rest-mass) and therefore we have added the subscript "0."

Now suppose I am moving with velocity v with respect to the system. Then the momentum is $p = vm_0$, and the energy, E, changes into the energy at rest m_0c^2 plus the kinetic energy $m_0v^2/2$ or $E = m_0c^2 + m_0v^2/2$. Then our equation is

$$(m_0c^2 + m_0v^2/2)^2 - (cvm_0)^2 = (m_0c^2)^2$$

Is this equality correct? To check it out, we must evaluate $(m_0c^2 + m_0v^2/2)^2$ and get $(m_0c^2)^2 + m_0^2v^4/4 + 2(m_0^2c^2v^2/2)$. The $(m_0c^2)^2$ term cancels with the similar term on the other side of the equation. The $c^2v^2m_0^2$ term cancels with the same term from the momentum. We are left with $m_0^2v^4/4$. The equation does not check.

Where did we go wrong? We went wrong by saying that the kinetic energy is $m_0v^2/2$. We also went wrong by setting $p = vm_0$. According to Einstein this is not true. Einstein proposed that for a moving observer, E and p are changing as t and v change. This is stated in a simple way by introducing a "mass" called m which differs from m_0 and is $m = E/c^2$. Then it turns out that the momentum is $p = mv$ and that m is greater than m_0 in the ratio

$$m = \frac{m_0}{\sqrt{1 - (v/c)^2}}$$

It is not difficult to see that, for v much smaller than c, all the statements of Newton remain correct—in good approximation. Of course, $m_0^2v^4/4$ is very small compared to the other terms, if v is very small compared to c. This is the normal situation and the error, in saying that the kinetic energy is equal to $m_0v^2/2$, is so small we cannot detect it. This is similar to the different observed values of t under usual conditions. Of course, if the velocity happened to be very large (that is, almost as large as c), then E would be large compared to m_0c^2 and E could be approximated by cp.

I like this description of energy and momentum because it is simple. It is simple in this sense, it unifies three conservation laws.

Assume that all velocities are small ($v \ll c$). Then energy and rest-mass differ only by a factor c^2. Conservation of mass and conservation of energy mean the same thing.

Assume, on the other hand, that conservation of energy and conservation of momentum turn out to be valid for one observer. Then the invariant means that these conservation laws hold for all observers. Each component of a four-dimensional vector is conserved. In a new coordinate system (i.e., for another observer), the new components of the vector are combinations of the old components, and thus they are conserved. The conservation law holds for the vector itself. One can express this also by stating that conservation of energy could not be universally valid without conservation of momentum, and vice versa.

So far we have extended Einstein's theory of relativity from its original purpose, to describe space and time, to a new purpose, to describe momentum and energy. But Einstein did much more. He did not merely redefine these relatively simple concepts which go back to Newton, he also managed to give to Newton's great conception of general gravitation a novel, a rational (or rather, to use his own expression) "geometric" basis. Instead of Newton's empirical $1/r^2$ law, Einstein engaged in a truly extravagant adventure of the mind. In the geometry of four-dimensional space-time, he explained gravitation as a necessary consequence of the idea that this four-dimensional space-time has curvature!

We have now obviously arrived at the most difficult part of our discussion. Before you decide to skip it, I must tell you its advantages. It is short. I don't really expect that you will understand what I am going to write. I think you should just try to get a feeling for what I am writing. The other advantage is that it is simple (for me). You shouldn't feel left out when I say, "for me." It is simple in the sense that it makes other things easy to understand. It is difficult because it is novel. Before you have understood general relativity (which is Einstein's name for the theory of gravitation), you are like the man who thinks the world is flat. After all, the world looks flat and if it were really round, why wouldn't people on the bottom fall off? To adjust from the obvious, that the world is flat, to the novel, that the world is round, is difficult, but it has its rewards. So it is with general relativity. It is novel, and therefore difficult, but its reward is greater simplicity in the end.

In describing gravitation, we noted that the force of gravity was proportional to the mass, and that the acceleration is obtained if you divide by the same mass. Therefore the gravitational acceleration is the same for all bodies. This statement holds with incredible accuracy. Why?

You are sitting reading this book. Suppose I tell you that the reason that you are pressed against the chair is not because of the earth's gravity, but because you have a rocket motor under you which is accelerating. That would mean that any object, if dropped, would seem to be accelerating in the same way, precisely canceling the acceleration produced by the motor. In fact, you cannot tell for sure whether you are being accelerated, or whether you are in a gravitational field, unless you look out the window and see that you are indeed at rest on earth. The idea that the effects of acceleration and the effects of a gravitational field are equivalent is called the "principle of equivalence." This principle of equivalence is at the heart of general relativity.

Unfortunately these accelerations seem simple only in a limited, local region. In a falling elevator, I can imagine that I am in free, gravitationless space. After a few seconds of enjoying this detached situation, I shall, however, suffer a shock. As an orbiting astronaut, I feel no acceleration (since the capsule and I are in the same orbit). I should imagine that I stand still or proceed in a straight line. Looking out the window will convince me that neither is the case. On earth, I here in the United States and my friend in Australia cannot imagine at the same time that the ground under our feet is accelerating upward—without assuming that the earth is on its way in a grand explosion. The problem is, therefore, how to reconcile the principle of equivalence which is most useful "in the small region" with a view of a coherent world that must hold "in the big region."

A similar situation exists in spherical geometry when one attempts to describe the surface of the earth in just two dimensions. In a small neighborhood, you may imagine that you move in a plane—the tangent plane. But considered as a whole, the surface has properties quite different from those of a plane. What shall we substitute for a straight line? It should be the line on which you move if you try not to change your direction. These should also be the shortest lines between pairs of points. On earth, this would be a great circle. Also these "straight lines" on the globe always intersect. There are, in this geometry, no parallel straight lines. (If, for instance, you follow lines of constant latitude, you do not move from one point to the other in the shortest distance.)

Gauss, the great mathematician, was deeply interested in these problems. He applied geometrical reasoning to all kinds of crooked surfaces. Then, in the beginning of the nineteenth century, a young mathematician, Riemann, who wanted to become a professor in Göttingen, offered to lecture on such surfaces in spaces of any number of dimensions. Gauss eagerly accepted. The lecture of Riemann became a classic. It also became the foundation of Einstein's "geometrical" explanation of gravity.

But first we must get some more understanding of a curved, two-dimensional geometry. Look at Figure 1. Suppose I took a vector at the north pole pointing south. I start with this vector on the north pole and move it directly south, always keeping it parallel to itself, pointed south. I reach the equator. Then I will move the vector along the equator for 6,000 miles, always keeping it parallel with itself, that is, still pointing south. Then I will move my vector directly north to the north pole (yes, still keeping it pointed south, always parallel to itself).

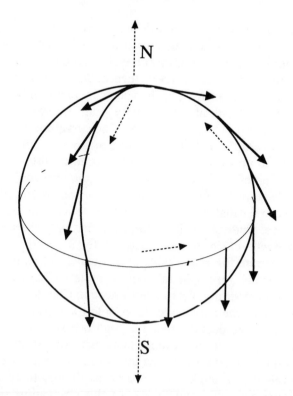

Figure 1.   Start at the north pole with a vector pointing south. Now take the journey given by the three dashed arrows: down to the equator, around it, then back north, being careful that the vectors are on the surface and always point to the south as you carry them on your journey.

Surprise! The vector has been turned during its journey in spite of the fact that I kept it parallel to itself all the time. The angle by which the vector was turned is proportional to the area that the vector has traveled around. This turning of the vector, divided by the area, tells "how much" the surface is curved. If the surface is sharply curved, the vector will be turned a lot and the "curvature" will be large. If the surface is gently curved, the vector will be turned a little and the curvature is small.

On a sphere the curvature will be constant everywhere. On other two-dimensional surfaces, the curvature need not be constant. To find the curvature, we can always travel around a little triangle and find how much the vector has been turned. So, for each little triangle, we can describe the turning of the vector by one number.

Now when I talk about relativity, I am interested in four dimensions, instead of two dimensions. The curvature in four dimensions is very complicated. If I go around an area in four dimensions, my vector can now turn in all sorts of directions, instead of just one direction. Therefore, in four dimensions, the curvature cannot be described by a single number. In four dimensions you need 20 numbers. You can see that it will be complicated, but do not be dismayed, you need not understand the complications, only the idea of curvature. We will return to these ideas shortly.

To illustrate the remarkable consequences of similar ideas in four dimensions, we might take an example which involves time. You will remember my trip to Andromeda in Chapter 1. I aged only one year during the trip, while the people I left on earth aged four million years. We decided that the only occasion when this time difference could occur is the event when I turn around at Andromeda. At that time, I undergo a strong acceleration, which according to the equivalence principle, is the same as though there was a big gravitational body underneath me. At this point, time on earth is passing much more quickly than time for me. On earth (according to my imagination of a great attractive mass under my feet) the gravitational potential should be high. It seems that high gravitational potential is connected with a fast passage of time, low gravitational potential with a slow passage of time. Ideas of this kind were involved when Einstein embarked on the puzzles of space-time curvature. It took him more than a decade before general relativity was completed.

The special name of the phenomenon we discussed above is the "gravitational redshift." If time passes more slowly, the electrons within atoms will oscillate with a smaller frequency and the characteristic spectral lines will be shifted toward the red. There are other reasons why a spectral line may shift. But if two stars travel together (a double star), the shift should be the same. If one of them is very dense, then on its surface the gravitational potential should be low and a redshift, as compared to the other star, can be observed.

It is amazing that exceedingly accurate measurements on earth established the redshift in Cambridge, Massachusetts. We have quantitative evidence that time passes at a different rate at the top and the bottom of the Harvard Tower.*

We should now relate the equivalence principle to the idea of curvature. Take two events which are simultaneous, but occur at slightly different places with different gravitational potentials. I want to take the vector composed of r, the distance between the two events, and t, the time between the two events, around a closed path. Actually instead of taking the vector around a closed path, I will take it on two different paths, both of which have the same initial and final point. This is equivalent to going around a closed path. The first path is: wait a while, and then displace the vector down one meter. The second path is: displace the vector down one meter and then wait a while. Because of redshift, the two "wait-a-whiles" are not going to yield the same result. The vectors displaced on two different paths will point in two different directions. In fact the time difference between the two end points of the vectors (which originally was zero, because we started from a pair of simultaneous events) will not be the same. Our four-dimensional space is curved.

We have found that gravitational phenomena can be described by lines of forces. We have further stated that these lines do not end, except in a region where there is some mass. It follows that in a region of empty space the number of lines entering the region and departing from it must be equal.† We cannot develop here the correlation between that picture and space curvature. But we can sketch the results.

---

* The difference is much too small to enter into regulations concerning the length of the academic year.

† We used this fact in Chapter 4 to derive the $1/r^2$ law.

Space is curved whenever a nonuniform gravitational field is present. That is, the curvature exists not only where there is mass (which causes the curvature) but also around it. The law that in free space the same number of lines enter and leave can be derived with moderate difficulty by discussing redshifts in a manner which we practiced above. From this the $1/r^2$ law follows, Einstein having used the hypothesis of equivalence (of acceleration and gravity) and the courage to consider the world not only being four-dimensional but also curved. The coupling of the curvature to the masses (and to related physical entities) was guessed by Einstein in a plausible and beautiful manner. No one has as yet found a way to improve his guess.

One example is a peculiar modification of Kepler's law: Mercury actually does not travel on an ellipse, it travels on a "rosette" orbit. Now I told you that from space curvature we could calculate the $1/r^2$ law. This law is an approximation which works in a certain reasonable distance from the sun. Mercury is the planet closest to the sun, and for this distance from the sun, space curvature predicts a slight modification of the law that explains Mercury's orbit.

Space curvature also explains how light should behave as it passes a heavy body. And indeed, light bends as it passes the sun, just as general relativity says it should.

In addition to these small effects, we also have a chance to understand truly spectacular events. I am talking about supernovae. A supernova is a star which flares up and gives off radiation 100 billion times as intensive as the sun. We believe that the interior of the star collapses and releases gravitational energy, which is transferred to the exterior of the star and becomes visible as light.* The first recorded supernova was observed in 1054 A.D. by the Chinese Imperial Astronomer. He kept exact records of position and brightness and so we know the location of that supernova. When we look at the "has-

---

* The light is a small fraction of the energy. Most of it is difficult (some of it almost impossible) to detect.

been" supernova, we see the Crab Nebula, which is a very large cloud. In the middle we see a very faint star, which we believe was the original star. This star is 10^{14} times as dense as water. It is a pulsar, which rotates 30 times per second. We want to study it. Because of its high density, Newtonian gravitation—which is only an approximation, good for weak fields—does not suffice. Einstein's theory should come into its own here.

We suspect that a supernova collapse may become so violent as to produce a black hole. In the zoology of stars, one may call a black hole a Boojum.* If you are attracted to it, you will "never be met with again."

You probably will not agree with me that all this is simple. Much of it I did not even dare to explain. But it reduces a part of nature—gravitation—to something conceptual and therefore inherently simple: geometry.†

Are you confused? If you are thinking about this material for the first, second, or even twentieth time, you should be confused. Confusion is not a bad thing. It is the first step toward understanding. If you are not confused by such a topic, it means you imagine that you understand it. You will not think more about the topic. If you are confused, you know you don't understand it. You cannot begin to understand until you realize that it is both interesting and rather tricky.

Are you unconsoled about your confusion? Let me tell you that Einstein received the Nobel Prize, not for his theory of relativity, but for a suggestion he made in 1905 (the year he started publishing about relativity). He argued that light, which had been considered as a wave motion, behaves as though it consisted of particles. For this additional puzzling but simple statement, he received the highest award a physicist can get.

---

* See *The Hunting of the Snark* by Lewis Carroll.
† ET does not fully admit that it is exceedingly messy geometry. Physicists assume that mathematics is a tool and therefore has no right to be difficult.

QUESTIONS

1. In 1054 the Chinese Imperial Astronomer did not have photographic plates or other equipment that astronomers now use to tell the brightness of stars. How do we today know the decay he observed in the brightness of the supernova?

2. Estimate the deflection of light by the sun, assuming that light consists of particles moving with velocity c and that the acceleration obeys the $1/r^2$ law. (Reality and also calculations based on space-curvature give a greater deflection by very nearly a factor of 2.)

Chapter 6

STATISTICAL MECHANICS

Disorder Is also a Law

*In which the reader finds out about heat and
perpetual motion of molecules, which,
however, never can be used to make
a perpetual motion machine.*

One of the important uses of the theory of atoms (or the combination of atoms called molecules) was to explain the phenomena of heat by the motion of the molecules. The point is that the motion of molecules need not be followed in detail. Statistics suffice. Hence the name "statistical mechanics."

The most obvious example is to explain the pressure of a gas as being caused by molecules bumping into a wall. It is clear that the more molecules you have, the greater the pressure. It is no surprise that pressure and density are proportional in the case of gases where the molecules bump individually and independently.

A second statement, that pressure and temperature are proportional, is not so obvious. The problem is that I haven't told you what temperature is. For a beginning, I say that heat is caused by the motion of molecules; the faster they run around, the hotter it is. Actually temperature is proportional to the kinetic energy of the motion of the molecules in the gas.* The kinetic energy is $mv^2/2$, where m is the mass and v is the velocity of a molecule. We said that pressure is caused by molecules bumping into a wall and giving the wall their momentum. The momentum is mv. The pressure must also be proportional to the number of times the molecules bump into the wall. The faster the molecules are travelling, the more frequently they will reach the wall. So the pressure must be proportional to the velocity of the molecules, times the momentum of the molecules, or $(mv) \times (v) = mv^2$. Thus it is plausible that pressure and kinetic energy should be proportional. The temperature is then defined to be proportional to pressure and, therefore, to the kinetic energy of the molecules or, rather, to the average value of the kinetic energy of the translational motion.

One sets the temperature equal to the average kinetic energy of the atoms in a gas, times a constant of proportionality. The interesting question is: Is that constant of proportionality the same for all kinds of molecules? For instance, oxygen molecules are 16 times as massive as hydrogen molecules. At a given temperature, do they move with one quarter the speed (on the average) as hydrogen molecules, so that mv^2 (essentially the kinetic energy and also the pressure) remains the same? They do!

This is connected with a simple, basic law of statistical mechanics: energies add, probabilities multiply. To understand this statement, look at two molecules colliding. The sums of their kinetic energies before and after the collision must be the same, because energy is conserved. The probability that the system consists of these

---

* WT: That is not how temperature was originally defined!

ET: Correct. It was defined by measurements using the volumes of gases. I did not tell you in the case of measuring time that it was originally defined by sand clocks and priest astronomers.

WT: But it was.

ET: The advantage of theory is that you do have to remember only a few facts and can claim that you know what you are talking about.

two collision partners must also remain the same before and after the collision, because the one follows from the other as cause and effect. But that probability is the product of the probabilities of the two molecules being in their respective states of energy.

Energies add and the total is conserved; probabilities multiply and the product is unchanged. This works only if there is a connection between energy and probability and, for a given temperature, this connection must be the same for molecules of different masses. If this were not so, then both energy and probability could not be conserved in the collisions between molecules of different masses. Temperature, of course, is not changed in the collisions. Indeed, temperature is in fact established as a result of many random collisions.*

As a slight and truly simple detour, it is nice to note that temperature has a lower limit. It does not have an upper limit; you can give molecules as much energy as you please. But we must specially consider the case when the molecule doesn't move. The kinetic energy is then zero, and so the temperature is what we call "absolute zero." You can't get any colder. On the Celsius scale ($0°C$ at the melting of ice, $100°C$ at the boiling of water), the absolute zero is $-273.15°C$, so that room temperature is about 300 K.†

---

* WT: What about the collision of earth and the meteor that made the meteor crater in Arizona? Was there no increase in temperature?
ET: Lots of increase.
WT: So?
ET: (On the defensive) I am talking only about collisions that occur in thermal equilibrium, *after* "temperature" is established.
WT: Are you begging the question?
ET: I am. All I am trying to say is that I can beg the question in a consistent way. Statistical mechanics consists in talking about a complicated system in a simple manner without contradicting oneself.
WT: Holy simplicity.
ET: We shall have more to say about the simplicity of Willard Gibbs, the only honest man in statistical mechanics.
† The capital K is the initial of Lord Kelvin, who introduced this "absolute" scale of temperature measurement.

CHAPTER 6

To sum up, I write the equation $P = \kappa Tn$, where P is the pressure, κ is a constant, T is the temperature (as measured from the "absolute zero"), and n is the number of molecules in one cubic centimeter (cm^3 or cc). This equation says nothing more than that pressure in a gas is proportional to the difference of the actual temperature and the "absolute zero" and to the density measured as the number of molecules per cm^3.

We must now address the way in which the probability of finding a molecule with a certain energy E depends on that energy. The probability is proportional to $P \sim e^{-E/kT}$, where the temperature T is measured from the absolute zero. At absolute zero the exponent becomes minus infinity and the probability becomes zero for any finite energy. At absolute zero the system of objects under consideration have their lowest possible energy. Actually, the exponential function we introduced is the only possible one where the energies (in the experiment) add while the exponentials multiply. We shall discuss its properties some more in connection with the way how the density of air depends on altitude.

This dependence of the probability on energy and temperature is called the Boltzmann factor and the constant of proportionality k is the Boltzmann constant. That k is the same for any molecule and for any state of matter is the reason for the simplicity and effectiveness of statistical mechanics.

All this can be made more obvious by an example. Let us study the distribution of the density of air as we go up toward the top of the atmosphere. I assume (and this is not quite true) that the temperature of the air is constant as you gain altitude. (Actually the air first gets colder and then, at very high altitudes, it gets much warmer. In fact, the variations of the temperature in the bulk of the air is relatively small—if we measure temperature in the correct wa, that is, from absolute zero.)

Now I want to consider a column of air between two altitudes, z and $z + dz$ (see Figure 1). The pressure at altitude z is higher than the pressure at altitude $z + dz$ because there is more air to support at z than at $z + dz$. The rate of change of pressure with altitude is proportional to the density of the air between the two altitudes. But

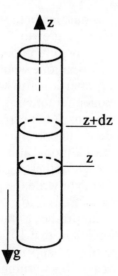

Figure 1. A column of air; note that the pressure is higher at z than at z + dz because gravity points downward.

we have assumed that the temperature is constant, so pressure and density are proportional. Thus we have $dP/dz = aP$, where a is a constant.

I want to change this equation in two ways. First, we know that if we go to a higher altitude, then the pressure drops, so I should have written $dP/dz = -aP$. Second, we can make the formula $dP/dz = -aP$ more precise. The change in pressure actually is the weight of the air in a unit area between the two altitudes. This weight is $gnmdz$, where g is the gravitational acceleration and m is the mass of a particle. So we can rewrite the formula as $dP = -gnmdz$ or $d(n\kappa T) = -gnmdz$.

We have assumed that κT does not change. So, $d(n\kappa T) = \kappa Tdn$. I want to get a little more for my work by obtaining in a clear-cut way how the density of particles n changes with the altitude z: $dn/dz = -(gm/\kappa T)n$. Since n and P differ only by a constant factor, it is equally true that $dP/dz = -(gm/\kappa T)P$. The equation says that the rate of change of the pressure is proportional to the pressure.

If we didn't have the minus sign in the equation, we would have a situation similar to population growth. The rate of change of population is proportional to the present population. With the negative sign in the equation, we get what one would call population decay. The rate of the decay is proportional to what is left. This kind of formula leads to an "exponential function" for the pressure. What an exponential function means, I can explain as follows: If I go up by an appropriate altitude z_0, the pressure drops by a factor of 10. Then if I go up by the same additional altitude z_0, the pressure drops by another factor of 10 or, altogether, by a factor of 100, and so I can continue upward. We might write the pressure as $P = $ constant $\times 10^{-(z/z_0)}$. (Indeed, as z goes from 0 to $z = z_0$, P drops by a factor of 10; if $z = 2z_0$, then P diminishes a hundredfold.) We must remember the rule of exponentials, that $10^x \times 10^y = 10^{(x+y)}$.

I must warn you that things happen faster than you expect with exponential growth or decay. This warning can be illustrated by the story about a legendary Persian Shah who wanted to reward one of his intellectual subjects for inventing the game of chess. The Shah was so delighted by the game that he offered the inventor any reward he desired. Instead of requesting gold or gems, the subject made the following, seemingly modest request: "On the first square of the board, put one grain of wheat. On the second square, put two grains. On the third, put four grains, and so on. Each time doubling the number of grains on the preceding square." This was exponential growth. The Shah said: "But, of course. Is that all you want?" But on the rth square the Shah had to give the subject a total amounting to $2^r - 1$ grains of wheat. Even with his large storehouses, he could not fulfill his subject's request of approximately 10^{21} grains of wheat, or around 10^{14} tons. As is so often the case with the best stories, the end has been forgotten. The intellectual subject was either executed or else obtained unlimited power. But the moral to the Shah and us is: Beware of exponentials!

In writing the formula for pressure, mathematicians use a number e instead of ten.* This number is a little less than three. (More

---

* The advantage of using the number e is that it solves the simple equation $dx/dz = x$ by setting $x = e^z$.

precisely, it is $e = 2.718 \cdot \cdot \cdot$.*) Our final result is that pressure is proportional to $e^{-gmz/\kappa T}$. Remember that mg is the gravitational force that acts on a molecule and gmz is the potential energy of the molecule so, in fact, we have the pressure proportional to $e^{-E_p/\kappa T}$, where E_p is the potential energy. This proportionality remains true under all kinds of conditions. We have, indeed, found a formula in which the temperature T plays an important role. The generalization of this formula is what we had already guessed on the basis of primitive arguments. We also found by comparison that our hypothetical constant κ is Boltzmann's constant, $\kappa = k$.

Now suppose I have a molecule with some potential energy. This molecule falls, changing its potential energy into kinetic energy. In the process, the number of molecules, or in other words the probability of finding such molecules, does not change. We should consider the quantity $e^{-E/kT}$, where E is the total energy. The quantity $e^{-E/kT}$ does not change unless the molecule collides with a partner. $e^{-E/kT}$ is a measure of the probability of finding the molecule. The bigger E is, the smaller $e^{-E/kT}$ is. This means the more energy a molecule has, the harder it is to find the molecule of that kind.

Indeed, $e^{-E/kT}$, the Boltzmann factor, is the essential and really interesting part in the statements of statistical mechanics. A complete expression for the probability of finding a part of a system with certain properties is the product of two factors. One of the factors is interesting and the other uninteresting. We have just described the interesting one—the Boltzmann factor. For a single atom, the uninteresting factor (called the *a priori* probability) says that the greater the space you look for the atom, the likelier you are to find it. And also, the greater range you allow for its possible momentum values (or velocities), the better your chances of finding it. The complete formula for finding part of the system depends on $dx \; dy$

---

* The number e has an infinite number of digits to the right of the decimal point. For those familiar with American history, the first nine digits after the decimal can be remembered by $e = 2.7(\text{Andrew Jackson})^2$, or $e = 2.718281828. \ldots$, because Andrew Jackson was elected President of the United States in 1828. For those good in mathematics, on the other hand, this is a good way to remember their American history.

$dz\,dp_x\,dp_y\,dp_z e^{-E/kT}$, where everything but the last factor is obvious, uninteresting, but still essential.*

All of this had been clarified by the end of the last century by Boltzmann. However, the scientist who gave the most clear-cut explanation was Josiah Willard Gibbs, a peculiar teacher of science. He had a remarkable respect for the bodily reality of his ideas. For example, he had no difficulty in showing on the blackboard how the pressure of a liquid depends on its density. But to show how the liquid behaves as density and pressure and temperature change, he required a three dimensional model. This he would sketch in empty air, using his hands. As he lectured, he would refer back to this (invisible) model, being careful not to walk through it or puncture it with a careless hand, as this would have destroyed its reality. Today, using the miracle of computers, Gibbs could have worn gloves with computer-coupled instruments that would automatically translate his hand motions into any desired projection on a computer screen and be viewed by the students using appropriate glasses as a model in three dimensions.†

In elucidating statistical mechanics, Gibbs used the same method, simple but abstract, that we employed at the beginning of this chapter to show that the interesting Boltzmann factor had to appear. The main argument that Gibbs presented was this: Suppose there are two systems, two molecules with energies E_1 and E_2. Then the total energy of the system $E_1 + E_2$, has to be conserved if the two molecules interact but are isolated from everything else. The joint probability of the two systems, on the other hand, has to be

---

* WT: Why call the factor $dx\,dy\,dz\,dp_x\,dp_y\,dp_z$ the *a priori* probability? Why not call it the "I knew it already probability"?

ET: Because I learned too much Latin as a youth. A further important reason is pointed out in the first pages of Willard Gibbs's famous book *Elementary Principles in Statistical Mechanics* (1902). If you consider a bunch of particles (e.g., molecules) moving independently under conditions where their energy is conserved, having coordinate and momentum values in the range x to $x + dx$, y to $y + dy$, z to $z + dz$, p_x to $p_x + dp_x$, p_y to $dp_y + dp_y$, p_z to $p_z + dp_z$, then the coordinates x, y, z will change, the momentum values p_x, p_y, p_z will change. But the product $dx\,dy\,dz\,dp_x\,dp_y\,dp_z$ will remain unchanged.

† Gibbs was only slightly crazy, being less than a century ahead of his time. Clearly, he was a poor second to Democritus, a gentleman we will meet in Chapter 8.

obtained by multiplying the two isolated probabilities. (The probability for two dice both showing a "1" is 1/36, that is, 1/6 times 1/6, the product of the probabilities of each die showing a "1.") Then the Boltzmann factor for the joint probability must be $e^{-(E_1+E_2)/kT} = e^{-(E_1)kT}e^{-(E_2)/kT}$.

Note that $E_1 + E_2$ is the total energy of the two molecules. If our two molecules collide they will have new energies E'_1 and E'_2 after the collision. But $E'_1 + E'_2$ equals $E_1 + E_2$, because the total energy must be conserved. This is most convenient, because we must have the same probability of finding the two molecules before their collision as after their collision. So Gibbs argued that the probability should depend on the energy in such a way that probabilities multiply but energies add. The only way to do this was to let probability depend on energy as an exponential function. The same holds not only for two molecules but for any two interacting systems. It was this argument that was the foundation of Gibbs's *Statistical Mechanics*.

Let me give three examples of changes in systems that can be explained on the basis of statistical mechanics: a fast change, a slow change, and a very slow change.

For the fast change example, consider a sudden increase (or decrease) in the concentration of a material. The region of change expands (or contracts) and communicates its behavior to its neighbor region, which in turn expands or contracts, and so on. The velocity with which the disturbance moves is called the sound velocity. In the air, it can easily be guessed that the sound velocity is similar to the average velocity of the molecules in the air. (This is not so in solids or liquids, because there the properties of atoms and molecules play more specific roles.) Room temperature, 300° Kelvin (K), means that the average energy of air molecules is $kT/2 = mv^2/2$, where $k = 1.37 \times 10^{-16}$ ergs/K and, with an average molecular weight of air as $m = 28.8/6.02 \times 10^{23}$, the average air molecule velocity will be approximately $v_{rms}* = 290$ meters/sec, whereas the velocity of sound

---

* ET: The subscript rms means root-mean-square, since we took the square root of the average square of the velocity. Sorry.
WT: You should be.

in air at normal temperature and pressure is 331 meters/sec. The important part of this calculation is the combination of constants k/m, because even if you don't know the diameter of the molecules, we see that knowing the sound velocity gives you k/m.

For the slow process, think about how long it takes to detect the presence of a quiet skunk (or the perfume on a fashionable lady). Here the potent molecules start from the source and reach our noses only after repeated collisions with the air molecules. At each collision the odorous molecules lose their sense of direction. They slowly diffuse, moving a step λ in length each time but only getting $\sqrt{n}\lambda$ away from the source after n collisions.* For air, $\lambda = 10^{-5}$ cm (about a thousand times the size of the molecules)† and the stinky molecules move at about 2×10^4 cm/sec. If so, they make $2 \times 10^4/10^{-5} = 2 \times 10^9$ collisions per second and in an hour the bulk of the molecules move only $(\sqrt{7.2 \times 10^{12}})\ 10^{-5}$ cm = 27 cm. Of course, some of the molecules, traveling more direct routes, reach our noses at much greater distances in much shorter time, with the help of a draft or the lady's fan. If this were not so, skunks would have had to develop other defenses.

The third example of a very slow process is a chemical reaction. (That is, the speed of a chemical reaction that is compared to the speed of a reaction out of control, such as an explosion.) Why do normal chemical reactions spread out much more slowly than the speed of sound? It is because the stable molecules do not approach

---

* If after N collisions the molecule has moved a distance L_N and if it is deflected at a right angle, its distance at the next collision will be $L_{N+1}^2 = L_N^2 + \lambda^2$. The right angle gives the average direction of deflection. L_{N+1}^2 could be smaller or larger with equal probabilities, so L^2 increases on the average by λ^2 in each collision. The result is $L^2 = n\lambda^2$ after n collisions.

† Noting that air when liquid is a thousand times denser than when a gas, Loschmidt, at the end of the nineteenth century, used the results we have mentioned concerning sound velocity and diffusion to arrive in an essentially correct way at the size of molecules. Indeed at a thousandfold density, the probability of collision should increase a thousandfold. The mean-free path should decrease to 10^{-8}, the approximate size of the molecule. History notes that he did not do it in a manner to convince skeptics such as Mach.

each other sufficiently close to react, except if they overcome a repulsive energy, called the activation energy E_a. Therefore, using the Boltzmann factor $e^{-E/kT}$, the reaction will go slowly. Chemists have a rule of thumb that if you raise the temperature by ten degrees celsius, a chemical reaction will occur in half the time. Indeed, if E_a/kT is about 20, so that e^{20} is 500 million and e^{-20} is 1/500 million. In such a case a reaction will occur, between gases having densities similar to air, in a little less than a second. Suppose that kT was taken at room temperature, that is, $T = 300$ K and we now go to 310 K. Then $E_a/kT = 19.4$ and the reaction rate is increased by $e^{0.7}$, approximately a factor of 2, in agreement with the chemists' rule of thumb. If E_a/kT is about 40, then $e^{-E_a/kT}$ is 1/500 million times 1/500 million, and such a reaction would take place over many years. In most of the reactions that chemists study, E_a/kT is between 20 and 40.

One last remark: Consider what happens when I drop a piece of chalk. The chalk's potential energy is changed into kinetic energy and when the chalk hits the ground, that kinetic energy is changed into the potential energy of the breakage and into vibrations of the fragments. Eventually these vibrations give rise to disorderly "thermal" motion. Now, can this process ever happen in reverse; will the fragments of the chalk ever compose themselves and jump back into my hand?

The answer is hardly ever, or rather HARDLY EVER, EVER, EVER. . . . In the case of the chalk, when it fell all the molecules fell in the same direction. When it hit the ground, all the energy went into vibrations in all directions, resulting eventually in a disorderly temperature motion. Order has turned into disorder and such a change is irreversible.

We have a new law: The state of disorder must always increase. This disorder can be measured, however it is usually measured in a funny manner only because scientists are lazy and would rather add instead of multiply. To get away with this, they use the "logarithm" which is the opposite of the exponential. Now disorder is a measure of probability. Probabilities should multiply. Increased disorder means increased probability. So "disorder" must multiply. The scientists take the logarithm of the disorder to get "entropy" and entropy has the happy quality that it adds instead of multiplies.

Now, for the first time we have a law which is not a conservation law. In our work during World War II, it was necessary to separate the uranium isotope U^{235} from the more common U^{238} isotope. Such separation can be dangerous and I was to find out how these operations could be made as safe as possible. I also had to report on my findings to a committee of military men which was headed by General Leslie R. Groves. I discussed the safety problem, interrupted periodically by an, "Are you sure, Doctor?" from General Groves. Finally a Colonel, who clearly didn't understand entropy, asked whether a situation in which all the U^{235} went accidentally in one direction and all the U^{238} went in another direction would not be dangerous.

Such a situation would be dangerous, I explained, but it was as likely as if all the oxygen molecules in the room crept under the table and left us to suffocate. Then General Groves murmured, "You admitted that it *is* possible!" I wanted to comment, rather impolitely, "If all your earlier questions were based on similar doubts, I should have been less worried." But Dr. Tolman, a highly respected scientist, came to my rescue and explained, "Dr. Teller means that it is, in practice, impossible." The General took my answer in good part and accepted my statements from then on.*

In fact, if General Groves could bring about spontaneous order making in the universe he would be a very rich man. He could, for example, solve any water shortage and, at the same time, solve the energy crisis. He would only have to take a large vat of water from, say, New York Harbor. Then he could coax all the fast moving particles to one side of the vat and get the slow moving particles to the other side. Where the fast moving particles resided, the water would boil and General Groves could use it for a steam engine. The slow moving particles would turn to ice and, if General Groves were

---

* While General Groves may have been a little slow to understand science, he was also quick to understand some facts of life, such as the occasional need for swift changes of mind. I went away from that table with greatly increased respect for General Groves.

sufficiently careful, he could get rid of any impurities and obtain potable water.

The idea of building a perpetual motion machine by separating fast and slow moving particles was first discussed by Maxwell, and the traffic cop directing the molecules was nicknamed Maxwell's Demon, in his honor. For years it was hotly (and sometimes unscientifically) debated why such a machine would not work; but such a machine contradicted the law that disorder always increases. It was a Hungarian physics student, Leo Szilárd, who gave the nicest solution of the problem. He pointed out that if the traffic cop managed to decrease the entropy, then the entropy of the cop himself would increase. In other words, Maxwell's Demon could separate the fast moving particles from the slow moving particles, but in doing it he would sweat. The information obtained by the Maxwell Demon must be included in a quantitative form in the calculations on statistical mechanics. Szilárd's paper (which was his doctoral dissertation) was the actual beginning of a science which is now called "Information Theory."*

Statistical mechanics is a funny subject since it talks about the behavior not of a single particle, but of an assembly of particles. A newcomer to the subject is tempted to say that statistical mechanics is interesting, but it is only statistics (that is, it is really not applicable to a small number of particles). So, the newcomer thinks: If I take ten particles (a statistician would certainly consider that too small a sample for reliable data), then I can play tricks by the help of a Maxwell's Demon. The fatal flaw in this argument is that as soon as you consider these particles over a long period of time, or consider many sets of a small number of particles, the laws of statistical mechanics take over. Even though statistical mechanics is based on averages, it is a theory that is universally applicable, provided you know how to apply it. In a certain sense, statistical mechanics is a

---

* Strictly speaking (and at that point rigor was necessary) Szilárd had to define the *minimum* entropy increase the cop had to accept if he was to direct molecular traffic.

precursor to quantum mechanics where we deal in a more fundamental way with probability and averages.

QUESTIONS

1. If an egg cooked to my taste takes 3 minutes to boil in San Francisco, but 6 minutes in Denver, and I know that the boiling water
at these locations has temperatures of 100°C and 90°C, respectively,
what is E_a/kT for the egg in San Francisco? (Remember that statistical
mechanics uses the absolute temperature scale and that absolute zero
is −273°C.) You must also assume that only one chemical process
is relevant in the egg boiling.

2. On a very hot day, does it make sense to open the refrigerator
door to cool down the house?

Chapter 7

ELECTRICITY AND
MAGNETISM
or
THE STRUCTURE
OF VACUUM

*In which it is shown that electricity and magnetism
are at least as intricately connected to each other
as space and time.*

I have tried to convince you of my philosophy that science is simplicity. Now, instead of making things more simple, I am going to make them more complicated. This is really part of the process of science. One understands something; it is simple. Then, one adds new phenomena and things become more complicated. One systematizes the new phenomena and things become simple again.*

As we will discuss in the next chapter, matter is made of molecules which, in turn, are made of atoms. The atoms are made of a heavy center, the nucleus, and light electrons. The nuclei and elec-

---

* WT: Sounds easy. ET forgot to warn you that you complete the trip back to simplicity only with a great deal of work and (in some cases) divine intervention (see Chapter 3). If the result of science is simplicity, its motive power is surprise.

trons are electrical in nature and their electrical nature holds them together to form an atom. Molecules are also held together by electrical forces. The fundamental nature of matter is electrical. In an explosion, the energy release is the result of a rearrangement of electrical forces. Electrical forces are so strong that normally the electrical charges are not isolated. Their effects are not easily observed and they were noticed only occasionally; they were systematized and understood in the nineteenth century.

Gravity was also not understood until the time of Newton, but for the opposite reason, it is very weak. The force with which you attract this book is much too small for you to notice unless you perform insanely careful measurements. By the time gravitational forces are large enough to be noticed, the objects involved are so big that one loses sight of what is involved.

The way in which electrical forces interact is similar to the way gravitating masses interact. If e_1 and e_2 are electrical charges, then the force between them is e_1e_2/r^2 where r is the distance between them. This is called Coulomb's Law. For masses m_1 and m_2, Newton's law of interaction is $-G(m_1m_2/r^2)$ where G is the gravitational constant and the minus sign indicates (by convention) an attraction.

These laws differ in two ways. First, the factor G does not appear in Coulomb's law. The factor G occurs because mass was defined on the basis of inertia and not on the basis of gravitational interaction. It is not an important difference. In fact, the electric charges are defined in such a way that no factor should be needed in Coulomb's Law.

The other difference is the minus $(-)$ sign in Newton's Law. This is not the case for electrical forces. There are two kinds of charges, positive and negative. If e_1 and e_2 are both positive or both negative, their product is positive and the resulting force, according to Coulomb's Law is a repulsion. If one charge is positive and one charge is negative the product is negative and the resulting force is an attraction.

In an atom or molecule, there is the same amount of positive charge and negative charge. These opposite charges hold onto each other so tightly that they hardly ever are separated and observed. Actually, they can be separated and the electricity can be observed.

But observations of events in which such separations occur and matter is "electrified" appear at first (and also at second and third) sight rather confusing. The Greeks had a word for it, but no explanation. That came two millennia later.

They also had a word for magnetism, but no inkling that electricity and magnetism are connected. In fact, the connection between these two phenomena provided the first step in modern physics beyond the Newtonian revolution.

One of the most easily observed effects is "static" electricity you can produce by combing your dry hair with a dry comb on a dry day. When you comb your hair, you transfer charge from your hair to the comb. Your hair will then carry a charge and, therefore, each strand of hair will repel its neighbor. The result will be that your hair stands on end so that each strand can get as far away from its neighbor as possible.

Another electrical effect, lightning, was, of course, noticed by the first man, but it was not known that it was electrical in nature. It was the American journalist, politician, and inventor Benjamin Franklin who, with a foolhardy experiment (carried out with utmost caution), showed that lightning was electrical. Fortunately, Franklin understood what he was doing even before he ran the experiment and lived to a ripe old age.

What produces lightning? Water droplets or ice crystals fall through the air and as they fall they transfer electrical charges to oxygen molecules and other constituents in the air. When enough electricity is rubbed off, then it jumps across a big distance to produce a "big spark." What actually happens in the discharge is that an electrical charge or an electron* is accelerated to so high a velocity that when it bumps into a harmless molecule, it manages to tear

---

* We shall have much more to say about electrons in later chapters.

another electron away. Now there are two electrons being accelerated and when they bump into another molecule you will obtain four electrons and so the process continues. You have an exponential growth of electrons, an avalanche. In this way, you get lots of electricity moving in one direction and this energy is released in light, heat, and sound. This is a mere outline of what happens in lightning. Franklin knew nothing of all this. He only suspected and subsequently proved that Zeus and Thor were really gods of electricity. What happens in real detail is not understood to this day.

Lightning moves through air. Air is an insulator, something that usually does not let electricity move through it. There are other substances called conductors through which electricity moves easily. The difference between an insulator and a conductor is great. Consider a telephone wire, which is nothing more than a conductor, like a copper wire. It is surrounded by an insulator. Your voice, when you use the telephone, is changed into electrical impulses and these impulses travel along the wire. These electrical impulses would rather travel along the conductor a thousand miles, than travel a tenth of an inch across the insulator.

In the days of Franklin, the only big effect of electricity was lightning. A couple of centuries ago, the power of electricity and the refinement of electronics was beyond anybody's imagination. Would even Benjamin Franklin have opened his ears to the statement, "Matter is held together by electrical forces"?

It was the Italian physicist Alessandro Volta who, in 1800, noticed the first evidence for a massive transfer of electricity. His discovery led to the clear connection between chemistry and electricity and to the understanding of the movement of electric charges through water.

Water is a poor conductor, but not a good insulator. It can be made into quite a good conductor by dissolving some salt (for instance, sodium chloride, written NaCl by chemists—"Na" for sodium and "Cl" for chlorine) in it. The important point is that it is not atoms or molecules that get dissolved, but "ions." An ion is an atom with an electrical charge. In this case, we get Na^+, a sodium atom

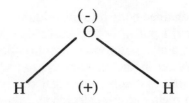

Figure 1.   The atoms in a water molecule are not arranged in a straight line.

with one too few electrons (called a positive ion*) and Cl⁻ (a negative ion) a chlorine atom with one too many electrons.† Why can we do this in water and not in air or another liquid like benzene?

Water is made up of two hydrogen atoms and one oxygen atom (H_2O, as the chemists would write it). Now, the atoms in a water molecule are not arranged in a straight line, but the molecule is bent as in Figure 1. It happens that the electrons from the hydrogen would rather hang around the oxygen than the hydrogens, so the molecule has one end with an excess negative charge and the other end with a positive charge. We call this a dipole because (a simplified model of) the molecule has one positive pole and one negative pole.

Now, it takes too much energy to decompose NaCl into Na⁺ and Cl⁻ if nothing else is around. But, if water is around, then the water dipoles cluster around Na⁺ presenting their negative ends to the ion while they surround Cl⁻ with their positive ends. This releases enough energy to permit the dissociation of NaCl into Na⁺ and Cl⁻ in solution. If I put a positively charged body called a positive electrode, into water with salt dissolved in it, then chlorine ions will be attracted to the electrode and sodium ions will be repelled by it. If

---

* Ion is a Greek word (invented by the British chemist, Michael Faraday) and it means something that is "on the go." Actually, it is "on the go" toward the electrode (the port facility of electricity leading into the solution) as we shall see later. (Incidentally, WT proposes to call an ion a "go-go atom.")

† To confuse matters, electrons have a negative charge so that Na⁺ has one too few electrons, despite what the "+" might seem to indicate.

I put a negatively charged body into the solution, the situation would be reversed. Now if I put in both a positively charged body and a negatively charged body, then sodium would collect at one body and chlorine would collect at the other body. Actually, I wouldn't get sodium and chlorine, but some chemical compounds made from them. If I were to melt sodium chloride and then put in a positive and a negative electrode, then I could separate the sodium and the chlorine. Using this method, magnesium chloride, found in the Dead Sea, can be separated and the valuable magnesium produced.

All of this was unknown to Volta. What he did find out, though, was that he could dissolve copper chloride in water and place a copper bar and a zinc bar in the solution. The zinc liked to be dissolved in the water, and Zn^{++} ions (that is, Zn atoms with two electrons missing) displaced the copper ions (Cu^{+}). The latter were pushed out of the solution and collected around the copper bar. Since the copper ions were positively charged, the copper bar became positively charged and since positively charged zinc ions were leaving the zinc bar, the zinc bar became negatively charged. If the two bars were connected by a metallic wire, electric current would flow. Actually, in a metallic wire (a typical conductor) only electrons move. These are shared by all the atoms in the metal and can move with little hindrance or "resistance." Franklin deflected considerable quantities of electricity; Volta caused them to move.

The real progress in understanding the situation was brought about by Michael Faraday, who suspected that his molecules were electrical in nature and who knew (due to the discovery by the Dane, Oersted) that a current (i.e., flowing electricity) produced an effect on magnets. He measured the electrical equivalent of different materials by depositing a certain amount of copper at one electrode and also determining how much electricity was transported in the process. Faraday did not measure the size of the copper atom, nor did he measure the charge of the electron; he measured the ratio of the two. Remember that Faraday was a chemist. Chemists of his time worked with the idea of an atom, although they never saw them. They did not know the weights of atoms, but they knew the ratios of the weights of different kinds of atoms. That Faraday applied similar ratios to electricity was new. It was a decisive step that was to lead toward simplicity.

Before the time when Oersted noticed with amazement that a magnet placed near a current was twisted, everybody imagined that forces acted only between centers of attraction or repulsion. Now, it was seen that a current is surrounded by magnetic forces. This led Faraday to his concept of lines of force. This concept can also be used to describe gravitation. We did so in Chapter 5.* We have seen that these lines illustrate the strength of a force at a certain place by the density of the lines and the direction of the force by the direction of the lines. This idea was invented by Faraday, but it was not immediately accepted by the contemporary physicists and mathematicians. After all, Faraday was a chemist.

Faraday also used lines of force to represent electric forces. The lines of force would originate in positive charges and end in negative charges. Of course, there is also another possibility that the lines would "close on themselves." An example of this last possibility occurs when the forces drive a current flowing along a loop of wire. This shows that when lines close on themselves, we have a source of power because a circulating current will release energy in the form of heat.

Faraday used these lines of force to invent and explain his Faraday Cage. A Faraday Cage is a box made from sheet metal.† If there are no charges inside the box, then no matter what electrical currents and charges are outside the box, there will be no static electrical forces inside the box. Why? Clearly, no electric lines of force can begin or end in the box. No line of force can close on itself in the box, because that would establish a source of power and we have assumed no source that can deliver energy inside the box. The only

---

* Newton, who was more of a mathematician than Faraday, did not happen to think of this simple tool which, in the end, has become an important part of mathematics.

† Actually, it is often made of wire mesh. Through this mesh, electric lines of force can "escape" only to short distances.

other possibility is that a line of force should enter the box from outside the box and then leave again. If this were the case, work could be obtained by moving a charge from one point on the surface to another. But, the surface is a conductor of electricity. The mobile charges present on the surface have already evened out any difference in the potential to do the work.

The only alternative left is the absence of electrical lines of force and, therefore, of electrical forces in the box. This has been fully verified.

The Faraday Cage is not only a simple and practical device. It is also a beautiful demonstration of the usefulness of a concept: The lines of force.

Oersted noticed that electrical currents had an effect on magnets. Faraday noticed that a magnet does not produce a current in a wire, but that a magnet which is moved causes electricity to flow. If a magnet is pushed back and forth through loops of wire, the electricity will flow first in one direction and then in the opposite direction, which is called an alternating current.* On the other hand, if there is an alternating current flowing through a wire, then a magnet will be pushed first in one direction and then in the opposite direction. With ingenuity, one can construct the wire in such a way that the magnet will rotate. This instrument is a dynamo, which turns electricity into mechanical energy.

Faraday constructed instruments so that by moving a magnet in one corner of the room he could make another magnet move in another corner of the room. In an embryonic form, he produced an electric generator and a dynamo. He used these instruments in his famous lectures for children, in which he would explain the most advanced scientific topics in simple terms, a practice that has, unfortunately, died out.

It happened that the then Prime Minister of England, Gladstone, attended the lecture where Faraday demonstrated his primitive gen-

---

* Only by adding the effect of moving magnets on electricity to the effect of moving charges on magnets was Oersted's discovery rounded out and made understandable. In the end, complete mathematical understanding was attained by Maxwell, who followed Faraday.

erator–dynamo. After the lecture, Gladstone went to Faraday and said, "This is all very amusing, but is there any use for this?"

Faraday said, "Sir, someday you may tax it." Of course, although Faraday did not do any engineering, what he had done became the basis for all the massive electrical engineering we use today. Any taxpayer should consider Faraday's response to Gladstone a beautiful British understatement.

There is a lesson in all of this. It is never clear when what appears to be pure science will become applied science.* It is also true that one never knows when progress in applied science will also benefit pure science.

After talking about how electricity induces magnetism and magnetism produces electricity, I should point out a basic difference between electricity and magnetism. Particles carrying an electron charge are well known (for example, the positive charge of a nucleus or the negative charge of an electron). Such a charge is surrounded by a radial, electric field obeying the $1/r^2$ law. A single magnetic charge is not known. Magnetic charges (if one wants to consider them at all) always come in compensating pairs so that the total magnetic charge is always zero. A magnet can be represented as the result of a current loop. In the case of the earth, there is such a loop, where the south pole is the positive end of the magnet and the north pole is the negative end of the magnet.[†]

Electric charges (for instance, electrons) do not produce magnetic fields, unless they are moving. Similarly, electric charges are not acted upon by magnetic fields unless they are moving. The magnetic field produced by an electron will be proportional to the velocity (v) of the electron and is perpendicular to the line of motion of the electron and to the line connecting the position of the electron with the point at which the magnetic field is measured. Similarly, the force that a magnetic field exerts upon an electron is proportional

---

* WT: Particularly not to prime ministers, presidents, or generals. Perhaps not even to bankers. The basic postulate of capitalism is, "From the fundamental facts stated above, there are more exceptions among bankers than among the other categories."

[†] Thus, the north pole attracts the positive end of the magnet in the compass.

to the velocity of the electron and is perpendicular both to the magnetic field and to the velocity of the electron.

If, for example, we have two charges e_1 and e_2 moving with the parallel velocities $\mathbf{v}_1$ and $\mathbf{v}_2$, respectively, the force of magnetic interaction between them will be $-(e_1 e_2 / r^2)(v_1 v_2 / c^2)\sin\theta$, where θ is the angle included between the velocity and the line connecting the two charges, c is the velocity of light, and the minus sign indicates that the magnetic interaction between parallel currents is an attraction. Actually, this attractive force is perpendicular to the velocities $\mathbf{v}_1$ and $\mathbf{v}_2$ and lies in the plane of the two velocities; it does not have the same direction as the electric interaction which points directly from one charge to the other.*

The speed of light seems to get into most basic equations in physics. In fact, one can measure the speed of light using the strength of magnetic fields. This is a provocative indication that light and electromagnetism are connected.

As I mentioned before, one of Faraday's most important contributions was the idea of lines of force. You can see these lines of force if you place a piece of paper over a magnet and sprinkle iron filings on the paper. The iron filings will form lines surrounding the magnet (see Figure 2). The jump of reasoning that Faraday made was that even when the iron filings were not present, the lines of force are still there. Lines of force can be used not only for magnetic fields, but for electric fields and gravitational fields.

All of these fields are forces that would act on the appropriate particles (charges, in the case of electric and magnetic fields, masses in the case of gravitational fields). They are qualities of space or

---

* WT: Are the two forces opposite?

ET: Yes. But, in general, they do not act along the line connecting the charges.

WT: What happens to conservation of angular momentum?

ET: One must ascribe some momentum to empty space. That is the peculiarity of Faraday's lines of forces. They carry energy and momentum. This was fully developed by Maxwell. Many physicists felt uncomfortable that empty space and lines of force should be so versatile. They tried to invent invisible wheels and mechanisms to explain electricity and magnetism. But the road to simplicity led in the direction of letting empty space and lines of force do the job.

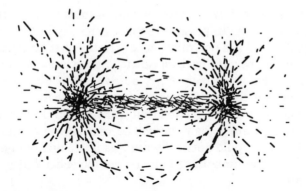

Figure 2.   This sketch illustrates what one might see if iron filings are scattered onto a piece of paper that covers a bar magnet; the filings appear to follow lines that run from one end of the magnet to the other.

qualities of the vacuum. You can see that vacuum isn't innocently filled with nothing, but rather, it is filled with potential forces: gravitational, electric, and magnetic. We have some reason to believe today that this is merely the beginning of a longer list. Unfortunately, we must defer the discussion of this list to beyond the last chapter.

The importance of the description by force fields is that we no longer accept the idea of "action at a distance": that would mean that something at one point produces a force at another point. Rather, a force is the property of space, each part of space having an influence on the adjacent region in space. This way of thinking was not completed until the discovery of the theory of relativity. The law that nothing moves faster than light holds for lines of force as well as for particles. If a magnet is moved at a certain time, then an associated line of force a light year away will not change until a year after the magnet was moved.

There is one additional point that should be made about the magnetic lines of force. If I draw any small box in a magnetic field, then the number of lines entering the box will be equal to the number of lines leaving the box. Any line that enters the box must either leave the box or end in the box, but it cannot end in the box because there are no isolated magnetic charges. Similarly, any line leaving the box must either have entered the box or must have begun in the box, the latter being impossible. Sophisticates say in this case that the magnetic field has no "divergence." Notice that an electric field can have divergence. If I have an isolated positive charge, then there will be lines of force which leave, but which do not enter a box surrounding that charge. If I have a negative charge in the box, there will be lines of force which enter, but which do not leave the box.

What I now am going to do will become a little difficult mathematically. You should not be discouraged if you do not understand all of it. It is more important that you get the general idea of what is happening.

What I want to do is to formulate laws which say that the magnetic fields of Faraday never start, never end, and try to encircle regions in which electricity is moving. In order to do this, I must have a convenient notation for the magnetic field.

I will write it as H_x, H_y, and H_z by which I mean that H_x is the component of the magnetic field in the x direction, H_y is the component of the magnetic field in the y direction, and H_z is the component of the magnetic field in the z direction. Remember, the component of the magnetic field changes from point to point. So H_x, H_y, and H_z will change from point to point.

By definition, the direction of the vector $\mathbf{H}$ is the direction of the lines of force and the size of $\mathbf{H}$ is the number of lines of force crossing a unit area that is perpendicular to the direction of $\mathbf{H}$. If we turn this unit area by an angle α, the number of lines crossing it will decrease by a factor $\cos\alpha$. The lines perpendicular to this area will also include an angle α with the line perpendicular to the original area, which was the direction of $\mathbf{H}$. But it is the property of $\mathbf{H}$ as a vector that its component including an angle α with the direction of $\mathbf{H}$ is $H\cos\alpha$. Therefore, if H is the number of lines of force per unit area of a perpendicular plane and if we call α the angle $\mathbf{H}$ includes with the x direction, then $H_x = H\cos\alpha$ is the number of lines of force crossing a unit plane perpendicular to x, H_y the number of lines of force crossing unit area perpendicular to the direction y, and H_z the number of lines of force crossing a unit area perpendicular to z.

Now we are ready to give the mathematical formulation of the statement that the number of lines of force that enter a volume must leave that volume. Consider a little volume element $dx\,dy\,dz$. By this, we mean a little rectangular box, a "brick," near a point with coordinates x, y, and z, whose six faces include: two parallel faces of area $dx\,dy$ at z and $z + dz$; two parallel faces of area $dy\,dz$ at x and $x + dx$; two parallel faces of area $dz\,dx$ at y and $y + dy$. Then the number of lines crossing the first pair of faces is $H_x\,dy\,dz$ and $[H_x + (\partial H_x/\partial x)\,dx]\,dy\,dz$. Here $\partial H_x/\partial x$ is the rate at which H_x increases if we proceed in the x direction. The ∂ (curly delta) is used instead of d to indicate that for ∂x only x is changing while y and z are held fixed. Similarly, the number of lines crossing other pairs of faces are $H_y\,dz\,dx$ and $[H_y + (\partial H_y/\partial y)\,dy]\,dz\,dx$ and finally $H_z\,dx\,dy$ and $[H_z + (\partial H_z/\partial z)\,dz]\,dx\,dy$. Taking the difference of lines of force through each pair (counting the departing lines minus those entering) gives $(\partial H_x/\partial x)\,dx\,dy\,dz$, $(\partial H_y/\partial y)\,dx\,dy\,dz$, and $(\partial H_z/\partial z)\,dx\,dy\,dz$. The total number of lines departing, minus those entering, will be the sum of the following three terms:

$$\left(\frac{\partial H_x}{\partial x} + \frac{\partial H_y}{\partial y} + \frac{\partial H_z}{\partial z}\right) dx\,dy\,dz$$

The factor $dx\,dy\,dz$ is the volume of our little brick, the volume element. The first factor is called the divergence of **H**. It is the important property of **H** that the divergence is zero:

$$\frac{\partial H_x}{\partial x} + \frac{\partial H_y}{\partial y} + \frac{\partial H_z}{\partial z} = 0$$

that is, the magnetic lines of force never start and never end.

The electric fields behave differently. They do take their beginning in positive charges and their end in negative charges. If we designate their components E_x, Ey, and E_z, then the corresponding relation is.

$$\left(\frac{\partial E_x}{\partial x} + \frac{\partial E_y}{\partial y} + \frac{\partial E_z}{\partial z}\right) dx\, dy\, dz = 4\pi\rho\, dx\, dy\, dz$$

The symbol ρ stands for the density of electric charges and $\rho\, dx\, dy\, dz$ is the total charge in the volume element. (The factor 4π is connected with our description of forces and lines of force. At the distance r from a charge e, the force due to this charge is e/r^2, so e/r^2 lines will cross outward through a unit area. Thus, at a radius r, the number of lines crossing will be the area of the sphere, $4\pi r^2$, times e/r^2, which gives $4\pi e$. Our equation containing the divergence is adjusted to conform to this formula.)

These, and the subsequent mathematical formulae were not due to Faraday, who was still a chemist,* but to his successor, James Clerk Maxwell, who was a theoretical physicist and who, therefore, did not have to see something in order to believe it.

The practical cause of magnetic fields are electric currents. Magnetic lines of forces occur in loops which encircle currents.† A simple model for the situation where a current in the z direction, i_z, gives rise to magnetic fields in the x and y direction, H_x and H_y, can be constructed by taking a little rectangle around the point x, y bordered by two pairs of parallel lines. (The z coordinate is left con-

---

* WT: ET has something about chemists. His father wanted him to become a chemist and ET was studying chemistry, but he wanted to study physics. His father, concerned (he was a lawyer, you see), took ET to Vienna to be interviewed by a relative, the famous Professor Ehrenfest. Ehrenfest asked the young ET, "Do you know what a curl is?" ET, with a glint in his eye, said, modestly, "Yes, sir." Accepting ET's word that he did, indeed, know what a curl is, Ehrenfest turned to my grandfather and said, "When I came to Vienna, I did know what a curl was, therefore I became a Professor. Edward already knows, so let the boy study physics." So ET became a physicist and we shall explain what a curl is.

† WT: What about bar magnets?
ET: They contain electrons which carry tiny circular currents even when they are at rest.
WT: Why, then, are not all substances magnetic?
ET: Because in most materials, electrons are paired in such a way that their currents cancel.
WT: Should I believe that?
ET: Yes.

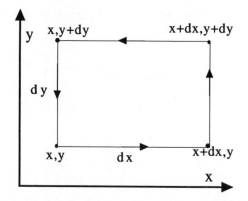

Figure 3.   Carrying one end of a magnetic dipole around the path shown enables us to relate changes (with position) of magnetic fields with electric currents and changes (with time) of electric fields.

stant and need not be considered.) One line has the length dx and connects the points (x, y) and $(x + dx, y)$. All four lines are shown in Figure 3. If we carry the positive end of a magnetic dipole in an anticlockwise direction along this line, the work done (force times distance) is $H_x \, dx$, provided the pole has unit strength. Continuing from $(x + dx, y)$ to $(x + dx, y + dy)$, one gets the contribution $[H_y + (\partial H_y/\partial x) \, dx] \, dy$ to the work, because the displacement from x to $x + dx$ changes the magnetic field from H_y to $H_y + (\partial H_y/\partial x) \, dx$. Moving from $(x + dx, y + dy)$ to $(x, y + dy)$, one obtains the contribution $-[H_x + (\partial H_x/\partial y) \, dy]$, and finally on the last lap from $(x, y + dy)$ to (x, y), $-H_y \, dy$. Adding the results of the first and third moves and canceling the common terms $H_x \, dx$, we are left with $-(\partial H_x/\partial y) \, dx \, dy$. Similarly, the sum of the second and fourth moves yields $(\partial H_y/\partial x) \, dx \, dy$. Adding these partial sums, we get the final result as $[(\partial H_y/\partial x) - (\partial H_x/\partial y)] \, dx \, dy$. The factor $dx \, dy$ is the area of the rectangle, while the first factor in the parentheses is a component of a vector called the curl of $\mathbf{H}$.

More completely, curl $\mathbf{H}$ is a vector whose z component is given, as we have seen, by $[(\partial H_y/\partial x) - (\partial H_x/\partial y)]$, while the x component may be obtained in an analogous way from the rectangle in the y-z plane and the y component from a rectangle in the z-x plane. Explicitly,

$$(\text{curl } \mathbf{H})_x = \left(\frac{\partial H_z}{\partial y} - \frac{\partial H_y}{\partial z} \right)$$

and

$$(\text{curl } \mathbf{H})_y = \left(\frac{\partial H_x}{\partial z} - \frac{\partial H_z}{\partial x} \right)$$

and, to repeat,

$$(\text{curl } \mathbf{H})_z = \left(\frac{\partial H_y}{\partial x} - \frac{\partial H_x}{\partial y} \right).$$

Now the use of the curl is that it is equal to $4\pi/c$ times* the vector of the electric current density with the components i_x, i_y, and i_z. That is,

$$\left(\frac{\partial H_z}{\partial y} - \frac{\partial H_y}{\partial z} \right) = 4\pi i_x/c$$

and, in similar fashion,

$$\left(\frac{\partial H_x}{\partial z} - \frac{\partial H_z}{\partial x} \right) = 4\pi i_y/c$$

---

* WT: Why 4π?
  ET: That is not simple. A factor of 2π comes from an integration along a circular magnetic line and a factor 2 comes from an integration along a straight line representing the current.

and

$$\left(\frac{\partial H_y}{\partial x} - \frac{\partial H_x}{\partial y}\right) = 4\pi i_z/c,$$

where the right-hand sides of the three equations are the total current flowing through the three little rectangles.

By writing down these equations and making them, through mathematics, more simple (an eternally dubious statement*), Maxwell managed to find something that eluded Faraday. The latter found that a changing magnetic field gave rise to loops, actually, to curls, in the electric field: curl $\mathbf{E} = -(1/c)(\partial \mathbf{H}/\partial t)$. To make this clearer, we write explicitly the z component of this vector equation:

$$\left(\frac{\partial E_y}{\partial x} - \frac{\partial E_x}{\partial y}\right) = -\frac{1}{c}\frac{\partial H_z}{\partial t}.$$

(The ∂ notation on the right-hand side means, of course, that we change the time but not the place.) Faraday looked, but could not find the relation with H and E interchanged. Maxwell showed that it had to be there, with the sign changed:

$$\text{curl } \mathbf{H} = \frac{1}{c}\left(4\pi\mathbf{i} + \frac{\partial \mathbf{E}}{\partial t}\right).$$

The first term in the parentheses gives the influence of the current, which we discussed. The second term is the influence on the magnetic field of a changing electric field, which Faraday could not find, because he could not measure magnetic fields accurately enough.

---

* WT: Glad you admit it. What is integration?
  ET: Integration is merely adding up ever so many pieces which get smaller and smaller.
  WT: I know. You try to explain it like differentiation, using ever smaller differences.
  ET: That is how people originally did it, yes.

Actually, there could be no question that the second term had to be present. What would happen to the equation curl $\mathbf{H} = (1/c)(4\pi\mathbf{i})$, if we had a finite current starting at one point of a line and ending in another? What would happen to the curl, the work on a magnetic pole? Around the line it would be present. If we leave the line, it would vanish. Would it change abruptly? One can show that the curl is a vector that must be represented by lines that have no beginning or end.

The combination $4\pi\mathbf{i} + (\partial\mathbf{E}/\partial t)$ accomplishes just that. If the current ends in a point, a charge must build up. But an increasing charge generates increasing electric fields. It is easy to show that $4\pi\mathbf{i} + (\partial\mathbf{E}/\partial t)$ has just the required property. The dependence of this kind of vector on position has the property that it can be represented by lines without end: where $4\pi\mathbf{i}$ stops, $\partial\mathbf{E}/\partial t$ will start.

But Maxwell did very much more. He showed that in the absence of charges or currents interesting processes can go on in the vacuum. The pair of equations

$$\text{curl } \mathbf{H} = \frac{1}{c}\left(\frac{\partial\mathbf{E}}{\partial t}\right), \qquad \text{curl } \mathbf{E} = -\frac{1}{c}\left(\frac{\partial\mathbf{H}}{\partial t}\right)$$

means* that magnetic and electric fields can induce each other and can become independent of material support. A most simple case is when the fields depend only on z and t, only the x component of $\mathbf{H}$ and the y component of $\mathbf{E}$ are present. Then the two equations become

$$\frac{\partial H_x}{\partial z} = \frac{1}{c}\frac{\partial E_y}{\partial t}, \qquad \frac{\partial E_y}{\partial z} = \frac{1}{c}\frac{\partial H_x}{\partial t}$$

---

* WT: What happened to $4\pi\mathbf{i}$?
  ET: We are in a vacuum, so $i = 0$.

For ct we may write τ, then the equations look simpler:

$$\frac{\partial H_x}{\partial z} = \frac{\partial E_y}{\partial \tau}, \qquad \frac{\partial E_y}{\partial z} = \frac{\partial H_x}{\partial \tau}$$

Two solutions to these equations are $H_x = E_y = f(z + \tau)$ and $H_x = -E_y = f(z - \tau)$. The first of these solutions is obvious. The letter f means any function. Changes of z and changes of τ have the same effect, so differentiation with regard to z and τ leads to the same result and the equations are satisfied. The electric and magnetic field remain the same if $(z + \tau) = (z + ct)$ is unchanged; if you wait a second and decrease z by c (by 186,000 miles) nothing is changed. We have an electromagnetic wave propagating in the negative z direction with the velocity of light. It is easy to see that the second solution $[H_x = -E_y = f(z - \tau)]$ is a wave propagating in the positive z direction. Such waves have wavelengths of miles (radio waves) or less than a micron (visible light) or even comparable to nuclear dimensions (as in cosmic rays).

One remarkable fact about electromagnetic theory is that, as developed by Maxwell, it is already in complete agreement with Einstein's relativity. An electric field at rest looks to a moving observer as a magnetic field, in part. One can see this relativistic behavior in a mathematical form by starting from electrostatics and using the electric potential ϕ, which is the potential energy of a unit electric charge. Then in relativity, ϕ becomes one component of a four-dimensional vector, like the energy which, together with the momentum, made up a vector of four components. In electromagnetism, one adds to ϕ, which is the potential for energy, a three dimensional vector with components A_x, A_y, and A_z which are the potentials for momentum. These four (ϕ, A_x, A_y, and A_z) behave like t, x, y, and z or like E, p_x, p_y, and p_z. Magnetic fields can be derived as $\mathbf{H} = $ curl $\mathbf{A}$ and electric fields as similar derivatives, for instance, $E_x = [\partial A_x / \partial(ct)] - (\partial \phi / \partial x)$.

Having talked a lot about fields we seem to have forgotten about matter. How do the fields interact with matter? This we have indicated in the first part of the chapter (electric fields exercise a force proportional to the charge in the direction of the field; magnetic fields give rise to a force proportional to the charge and its velocity and perpendicular to the field and the velocity). All this can be written in a relativistic form, which also has the great advantage of showing that the conservation of energy and momentum continue to hold in electromagnetic interactions.

In order to do it (and we shall merely indicate the procedure), we have to attribute some energy and momentum carried by the fields in space, even in the absence of matter. This is obvious if we remember that light is not material and that it carries energy in the amount $(1/8\pi)(E^2 + H^2)$ in a unit of volume. The complete story requires the introduction of a tensor, called the energy-momentum-tension tensor, which we shall not describe here.* I merely want to emphasize that we must attribute to space not only energy and momentum, but also a tension (described by a tensor in three-dimensional space and called the Maxwell tensor), which is similar to tensions in solids. The difference between matter and "empty" space seems to diminish in a remarkable fashion.

---

* ET: In three dimensional space we have seen that a vector has three components. For instance, the numbers F_x, F_y, and F_z, are the components of a force in three dimensions. Or dx, dy, and dz are the components of the displacement in three dimensions. A tensor may be formed from two vectors such that it has three times three, or nine components in three dimensional space.

WT: But from where did the model of a "tensor" come?

ET: The original use of a tensor was to describe tensions in a solid. Its components give the answer to the problem: If you cut a solid that is being squeezed or twisted and expose a small surface, you relieve a tension; what is the force needed to replace that tension? Actually, the components are many because you can cut the solid in many different ways and the forces themselves have components.

WT: And in four dimensions?

ET: The electric and magnetic fields *together* make up the components of a single tensor. Maxwell invented another one that gives the densities and fluxes of energy and momentum.

WT: And what is the advantage?

ET: Like old age, it is better than any other alternative.

WT: Thank you for not going any further with this discussion.

There is a story about electricity, energy, and the law. It happened that electricity was generated in North Carolina* and used to turn a motor in South Carolina. The federal government wanted to show that Federal Interstate Commerce laws applied to this situation, since something had been transferred from one state to another. The power company, wishing to show that federal laws did not apply, asked, "What have we taken from one state to another? We haven't taken electricity from North Carolina to South Carolina because if we had, North Carolina would have a tremendous charge." The government argued that energy was taken from one state to another state.

"Alright," said the other side, "if we have taken energy, show us where it flowed. Does it flow through the wire? Does it flow in the space around the wire?"

At this point, the judge got confused. He should have studied the energy-momentum-tension tensor, but there is no legal precedent to recognize that tensor in our books of law. In the end, I think he did the right thing. He ruled that money was transferred from state to state.† It must have been paid for energy, and the judge may have had a sneaking suspicion that energy is somehow conserved.

We have started this chapter with particles. We ended with fields. This corresponds to a great historic change introduced by Faraday and Maxwell. In the previous chapter, we started to replace (at least in part) rigid laws of cause and effect by statistical laws. The confluence of these two innovations of the last century, field theory and statistics, gave rise to the theory of atoms and of matter.

But first, we must raise the question: Do atoms really exist?

---

* The location may not be accurate, but the essential story is true. According to the great Hungarian Theodore von Karman, one must not be hampered in telling a true story by the accidental circumstances.

† Money is like electrical energy: it flows but one cannot localize how it flows. (That it flows from pocket to pocket is an obsolete superstition.) But in the case of money the law is more apt to recognize the invisible flow than in the case of electrical energy.

QUESTIONS

1. Charged bodies are attracted to a plane metallic surface. Why? How strongly?

2. Find a set of ϕ, **A** values compatible with $\mathbf{E} = \mathbf{H} = 0$ everywhere.

Chapter 8

THE EXISTENCE OF ATOMS

A Stands For Atom.
It Is So Small,
No One Has Ever
Seen It At All.*

*In which the reader will be convinced of something
he thought he knew. He will also find out that
atoms are not so terribly small.*

Perhaps you consider it strange that I should devote a chapter to the existence of atoms. It is obvious that I think atoms exist and you probably think they exist, too. But, why do you think so? Nobody has ever seen an atom with their own eyes. You've been brainwashed into believing in atoms. This belief is a beautiful example of today's "common sense."

The first attempt to convince people that atoms exist was made 400 years before Christ. The man who suggested it, Democritus,

---

* Written in 1945 by the same obscure Hungarian poet quoted in the Prologue. But on this occasion, time has proved him wrong. Atoms have been seen using the electron microscope and even better seen in the scanning tunneling microscope.

came from the wild northern part of Greece called Thessaly. He was invited to give a seminar on his ideas in Plato's academy. Democritus said that you can make things smaller and smaller, but there must be a limit. What was the evidence? Take water. You can evaporate it and it seems to vanish. Then, you may condense the vapor to get water. You can freeze water and then melt it and again you get water. Something must stay constant so that you can get the original material back. If you imagine particles of which water is made (what we call water molecules was called atoms, "indivisibles" by Democritus), then you have an explanation.

Plato was a conservative. He didn't believe that the earth moves around the sun even though Pythagoras hinted at it. He certainly wasn't going to believe Democritus's peculiar theories about matter. But, Plato was also compassionate. He was concerned about Democritus's sanity and so he sent the father of the atoms to the father of medicine, Hippocrates.*

The clinic of Hippocrates on the island of Kos was really quite modern. The discoveries of Freud (so the story goes) were anticipated by more than two millenia. Democritus was subjected to a double psychoanalytic session (that is, two times 50 minutes). The two Greeks emerged arm in arm and Hippocrates's verdict was, "If this man is crazy, so am I."

Unfortunately, the certified sanity of Democritus did not help. Academic doubts prevailed and the theory of Democritus was forgotten for 2000 years. The alchemists took over. They proposed the "scientific" theory of the four elements (earth, water, air, and fire, with ether as the excluded fifth)† and tried to change everything into everything else, especially into gold. The idea that one kind of matter could be changed into another kind of matter is something with

---

* ET brought this story home from Athens, specifically from the Greek Atomic Energy Laboratory called "Democritus." The people in that laboratory should know.

† WT: We have made much progress since then. Where the alchemists recognized earth, water, air and fire, physicists today deal with solids, liquids, gases and plasmas and have given up on the ether. This is progress.

which Democritus would have disagreed, because his atoms were basic and immutable.

Then around 1800, atomic theory emerged in a stronger form. The distinction between atoms and molecules was made for the first time by Lavoisier, a French nobleman who lost his head during the French Revolution.* The revived atomic theory established the important difference between what could change in chemical reactions and what was conserved, which are the hundred elements.

From this grew chemistry. For a hundred years, chemists built up a system of formulae. They could describe and predict reactions. They assigned detailed structures to molecules and positions to atoms within molecules. But, nobody had seen an atom. If they existed, they must be smaller than the wavelength of light, so that light waves could sweep over them as an ocean wave over a pebble.

The indications that the atomic theory was correct were strong, but indirect. Elements often combined in fixed proportion with each other. Hydrogen (almost) always combined with oxygen in the ratio 2 to 1 giving H_2O (which Democritus called an atom). Sometimes they combined in several proportions which had simple numerical relations to each other. For instance, one carbon could combine with one oxygen to give poisonous carbon monoxide (CO) or with two oxygens to give harmless carbon dioxide (CO_2). Hydrogen and chlorine always combine in the same proportion (HCl). If there was too much hydrogen available, then at the end of the reaction hydrogen would be left over and if there was too much chlorine, then chlorine would be left over. From the chemical combination, hydrogen chloride, the original hydrogen and the chlorine could be retrieved by dissolving HCl in water and passing an electric current through the hydrogen chloride.

All this was not as simple as it sounds. For instance, in H_2O, the weight of hydrogen is not twice the weight of oxygen. The ratio

---

* See the French play *Le Jeu de l'Amour et de le Mort,* by Romain Rolland (1924), for the story of Lavoisier. (The English translation was done in 1926: *The Game of Love and Death.*)

of weights is 1 to 8 rather than 2 to 1, because oxygen is (approximately) 16 times heavier than hydrogen. It was actually observed that the volumes of hydrogen gas to oxygen gas (at equal temperature and pressure) react in the ratio of two to one. This does correspond to the formula H_2O *assuming* that a given volume at a given temperature and pressure has always the same number of atoms, whether they are hydrogen or oxygen. Actually, according to statistical mechanics, the number of *molecules* are equal. But, how many atoms are in a molecule of hydrogen or oxygen? The answer happens to be two in both cases. It took quite a few years to straighten all this out. This was the first kind of argument that had to be tackled. In the end, it led to order in chemistry and it was a real success for the atomic theory.

Another kind of evidence which is usually less emphasized is the appearance of crystals. Consider, for instance, sodium chloride (NaCl) which you know as salt. It forms a lattice. Actually, it is the simplest common lattice. In each layer of the lattice, the sodium and chlorine atoms make a checkerboard pattern, alternating sodium and chlorine atoms as in Figure 1. The layer above (and below) this layer will be exactly the same except that everything is shifted so that above (and below) each sodium atom is a chlorine atom and above (and below) each chlorine atom is a sodium atom.

You say, "Why should I believe your statement about the pattern of salt? Can you prove it?" On the basis of the knowledge that was available in the early 19th century I can't, but I can start a proof. Take a crystal of salt and break it. It will break into little cubes, or at least into pieces with mutually perpendicular faces. Other crystals have other shapes and when you break them they also will split into their characteristic shape. These regular and reproducible shapes of crystals imply that they are made of atoms held together in a very regular order. It is evidence, but not proof.

At the end of the 19th century, two scientists, Ostwald, a German physical chemist and Mach, a Viennese physicist, who was the first to introduce relativistic ideas, challenged the atomic theory. They said, "Chemistry and crystals behave 'as though' there were atoms. So, let's stick to the facts, atoms don't exist. We may think about

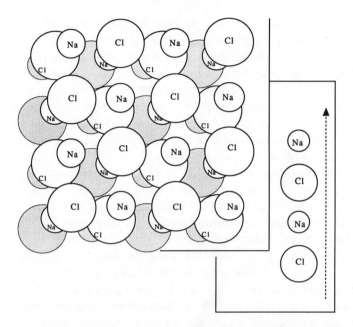

Figure 1. In salt, the sodium and chlorine atoms form a "checkerboard" array in each layer (above), with the sodium and chlorine atoms alternating from layer to layer (right).

them as infinitely small particles, enjoying themselves in the never–never land beyond human imagination, indeed, beyond the reach of any human experience. Only the ratio of their weights is real and only their mutual relation as described in the book of chemistry is verifiable. To believe that they really exist and are indivisible contradicts common sense (shades of Plato). To believe that they exist and are divisible deprives them of the proper purpose for which they were invented."

Ostwald and Mach were wrong, though they saw correctly the great difficulties atomic theory would generate. They also performed a great service: they got people mad. A great many experiments were devised. In the end, these experiments proved that atoms exist.

Atoms couldn't be seen, so how can one prove they exist? The approach that was taken was to assume they exist and then measure their size. Some methods were approximate, some were more accurate, but the important point was that of the dozens of measurements that were invented in unrelated fields, all led to the same size for the atoms. This induced people to finally accept the atomic theory. All this experimentation took place in a relatively short time. In 1900, people had no proof of the existence of atoms. By 1910, the existence of atoms was scientific fact.

Let us review some of the methods of measuring the size of atoms. There is a straightforward and crude experiment you can perform with things you have around the house. Consider waves in water. They behave differently depending whether they are long waves or short waves. The propagation velocity of the waves depends on the wavelength. (Note that this is not the case with light or sound waves which propagate with the same velocity independent of the wavelength. The velocity does, however, depend on the matter through which they travel.) Furthermore, in water even the type of velocity dependence is different for long and short waves. For long waves, the velocity increases with increasing wavelength; for short waves, the reverse is true: the velocity increases with decreasing wavelength. The reason is that essentially two different mechanisms are at play. The long waves in water travel because of the restoring effects of gravity. Gravity does not like the circumstance that at some places there is more water (at crests) and other places, there is less water (troughs). Gravity tries to equalize the situation and as a result the wave moves forward. Short waves in water are driven by surface tension. Water molecules practice togetherness. Water droplets are spherical because the sphere has the smallest surface area and so more molecules can stick happily together. A wave in water increases the surface area and the surface tension tries to rectify the situation causing the short waves to propagate.*

---

* ET: There is actually a way that Democritus could have guessed the size of the "atoms." He could have noticed that short waves (about a quarter of an inch long) travel a few inches a second on the surface of the water. Then he could have tried to excite waves

Soap has a very strange structure. One end of the soap molecule loves water and the other end hates it.* If I drop soap into water, it will orient itself so that the hydrophobic (water-fearing) end will be as far from the water as possible. Thanks to the gymnastics of the soap molecules, the surface tension will be reduced and a surface wave will travel more slowly.

Let us start short waves in the water by letting droplets hit the surface. The only apparatus you need is a leaky faucet (though a tuning fork struck and dipped into water works *much* better). If the surface is soapy, short waves will travel more slowly, which one can see when the waves hit a soapy patch by observing how the wave front is distorted.

Now, let us try to use a grain of soap to spread a film over a big area. It is surprising how much territory a little soap can cover (see Question 3 at the end of this chapter). But there is a limit. In the end, I will have put a monomolecular layer of soap on the surface of the water. If I measure the amount of oil† that produces a mono-molecular layer, then I can deduce the diameter of a molecule of oil. This, of course, gives a very crude value for the size of the mol-ecule. Furthermore, it is more difficult than it sounds. It requires clean water surfaces. But, if all this does not amount to a good ex-

---

by faster vibrations. (Pythagoras experimented with such vibrations before Democritus.) He might have found that waves that are tenfold shorter travel three times faster. (More accurately, as the square root of ten.) He could have imagined waves a hundred million times shorter and found that such waves should have moved with the speed of sound in water. Even shorter waves should have exceeded the speed of sound. That could not have been the case. A droplet of water a hundred millionth of an inch in diameter could no longer have the same properties as bigger droplets. This could give a hint about the divisibility of water.

WT: That sounds quite un-Greek to me.

ET: It's hard to imagine Democritus doing it. But you could do it with practically no equipment.

* That is why soap works. That other side loves dirt. It latches on to it. Then, the water-loving (hydrophilic) end anchored in the water, the other end sticks out of the water and the dirt is dragged to the surface.

† Oil can be used more simply because it is easier to get a well-defined monomolecular layer.

periment, it does give a wonderful demonstration. That there is a lower limit to the thickness of the oil layer is obviously related to the idea that one encounters difficulties when one attempts to subdivide molecules.

You may wonder why I measured the diameter of a molecule rather than the diameter of an atom. From the size of a molecule, I can calculate the size of an atom and it is easier to find the size of a molecule because they are both bigger and more available. In most of the experiments I describe, I will find the size of a molecule rather than the size of an atom.

Another method for determining the size of a molecule was inspired by a very old discovery of a botanist, Brown. The discovery was named, appropriately enough, Brownian motion. Brown noticed that very small particles, suspended in a liquid (he could see them under a microscope) moved. He concluded that the particles must be alive, which was a lovely idea but was, unfortunately, wrong. What was happening was that the particles behaved like large molecules. They were kicked from all sides, like molecules, so they shouldn't have moved. But there were fluctuations and these little fluctuations caused the particles to move slowly.

In 1905, the same year that he published his paper on relativity, Einstein deduced from Brownian motion the size of the molecules. Like all molecules, these particles had the same kinetic energy, $mv^2/2$, as did the other molecules. But since m is large, v must be small, although quite visible under the microscope. Comparing v (à la Brown) with the sound velocity, Einstein could get the mass of the molecules of air. His result was not so different from that of Loschmidt.

A Frenchman, Jean Baptiste Perrin, followed a similar path as Einstein. But while Einstein considered the kinetic energy of a particle, Perrin concentrated on the potential energy in the field of gravity. He put very small particles in water. Perrin noticed that the particles were distributed in the test tube according to the same law as the air above us. The number of particles per unit volume is proportional to $e^{-(mgz/kT)}$, where m is the mass of the particles, g is the gravitational acceleration, z is the altitude, k is Boltzmann's constant, and T is the temperature. Notice that if m is increased a millionfold, a z value, one millionth of what makes a difference in the atmosphere, will now count. Perrin's giant "molecules," actually called colloidal particles, produced a model of the atmosphere in a test tube. Perrin had to take the buoyancy of the particles in the water into account. The effective weight is then the weight of the particle minus the weight of the water the particle displaces. It is true that each of the particles was made of millions of molecules, but the barometric law held in the test tube. Perrin could determine kT and so he could calculate the mass of a molecule from the actual barometric law in air.

Like Einstein, Perrin obtained the ratio of the known mass of his particles to the unknown mass of the molecules in air. Loschmidt, Einstein, and Perrin all used the same principle: the kinetic theory of gases. An entirely different determination of the size of an atom shall be discussed in the next paragraphs.

I must introduce a new idea, the idea of interference. Suppose I have two sources for a wave motion. These two sources can be timed so that the crests of the waves arrive at the same time and so make an even larger crest. This reinforcement is called positive interference. I could also time my sources so that a crest from one source comes just when the trough of the other source arrives. Then, the waves would cancel each other and this is called "negative interference." We have already said that light has too large a wavelength to notice the atom. However, x rays have a much shorter wavelength. The German physicist Max von Laue decided to use x rays to see atoms. There were no x-ray microscopes and to make one would be very difficult.* Von Laue let x rays fall on a crystal. Crystals are lattices of atoms. The atoms scattered the x rays and in certain sharply defined directions the x rays underwent positive interference; in all other directions the x rays underwent negative interference. From

---

* It would also be extremely useful. We could see the structure of the molecules which transmit properties from one generation to another. It is difficult to make x-ray microscopes, but we are beginning to make x-ray lasers. With their help, x-ray microscopy will become easier. (We shall discuss lasers in the last chapter.)

this, von Laue established a relation between the distance between atoms and the wave length of x rays.

Unfortunately, we still have to find the wavelength of x rays. In order to find the wavelength, one can use gratings which are nothing more than parallel scratches on a plate which are very close together, say, ten scratches per millimeter. If x rays or light fall perpendicularly on a grating, they reflect back in the same direction. If x rays or light fall on the grating at an angle, they will be reflected at the same angle, because then light has to travel the same distance whether it is scattered by one scratch on the plate or its neighbor, and so the interference will be positive. Light also may leave at a somewhat different angle so that light has to travel one extra wavelength when reflected from a neighboring scratch. This gives a pattern of reflection called the diffraction pattern. Unfortunately for x rays, the angles between the rays in the diffraction pattern are so small (i.e., so small a change in angle will establish a path difference of one wavelength) that the structure cannot be seen. We have gotten no further in finding the wavelength of x rays. The happy surprise is that if the x rays impinge at a very grazing angle, less than one degree, then the angles in the diffraction pattern become great enough to be seen and measured. Now, we can calculate the wavelength of x rays in terms of the grating and so one can determine the size of atoms.

The next experiment clinched in most people's minds the argument. It was Millikan's oil drop experiment. If very small oil drops fall through air, they move slowly because of the friction of the air. By measuring the velocity of the oil drop, the size of the oil drop could be determined. Now, Millikan exposed the droplets to a gentle electric discharge so that electrons could attach themselves to the droplets. The charged droplets would fall with the same velocity. But, if they fell through an electric field, their motion would be affected. By noting the change of motion, Millikan could determine the amount of charge an oil droplet carried. It turns out that it could carry one of several charges, but they were multiples of one elementary charge which is the charge of the electron. Faraday had measured the electrochemical equivalent, that is, for example, how much electricity is needed to collect one gram of copper atoms. Knowing this and knowing the charge of an electron, Millikan could calculate the mass of a copper atom. Again, the same result was obtained for the size of the atom.

The next example which I want to describe is the simplest. You have heard of radioactivity. It is the emission of radiation: α rays, β rays, and γ rays. I need to discuss only the α rays. They are the same as nuclei of helium atoms. When an α ray is emitted, it has a high velocity and a lot of energy. If it bumps into an appropriately sensitive screen, light will be emitted. One can see a scintillation. Early workers ruined their eyes by counting scintillations. When an α ray finally comes to rest, it will collect two electrons and become a helium atom. Uranium will emit α rays and the number of α rays emitted in a certain period of time can be counted by counting the scintillations. Now, if I take a big piece of uranium and collect the α particles over a considerable length of time, I get a measurable amount of helium gas. I have counted the scintillations due to a small amount of uranium for a short time. I have collected helium from a much greater quantity of uranium over a longer time. So, in effect, I have counted the atoms emitted in the uranium decay. Conceptually, the method could not be more simple. I know how many scintillations have occurred, so I know how many helium atoms I have. I can calculate the size of a helium atom. The results agree again.

Now, at last, I want to contradict my first statement: molecules can be seen, not using a common microscope but using an electron microscope. The electron beam is trained on an object. The electrons which are deflected will then be focussed by magnetic fields at some other point and we see an image. Using the electron microscope, we can actually see molecules and determine their size. If I know the size of a molecule, I can determine the size of an atom. The results agree again.

Now that there is complete agreement of the size of the atom, the time has come to state the size. One hydrogen atom weighs 1.65×10^{-24} grams. The standard number of atoms that chemists use is called one mole and is the number of O^{16} isotopes (i.e., the abundant kind of oxygen) weighing 16 grams. This number is 6.02×10^{23} and is called Avogadro's number. The yardstick of atoms is the angstrom unit (Å) which is 10^{-8} cm.

After all, atoms are not so very small and the world of atoms is surprisingly accessible. A dog can smell a single molecule, provided the molecule is particularly odoriferous. Our eye can see a single light quantum, which we shall discuss later and which can be called an atom of light. I believe that when life will be understood, the understanding will be thoroughly connected with the theory of atoms. To that extent I am a materialist.*

We have disproved all the doubts of all the doubters starting with the Greek philosopher Plato, and ending with the Viennese philosopher Mach. We have seen the atoms. We cannot escape them. When you learn how they behave, you may wish that they did not exist.

QUESTIONS

1. Prove that the velocity of water waves changes as $\sqrt{\lambda}$ for long λ (λ is the wavelength) and as $1/\sqrt{\lambda}$ for short λ.

2. How many times per second does a molecule in air collide?

3. How large an area would a bar of soap cover if spread out in a monomolecular layer?

Chapter 9

THE CORRESPONDENCE PRINCIPLE

A New Science Based on a Contradiction

In which Niels Bohr does what has never been done before. Instead of laying down the law connecting atomic theory with conventional physics, he establishes the rules of coexistence.

The idea that scientific theories are based on observations is not always true. In 1930, a friend of Niels Bohr, the Danish scientist who changed the nature of science, asked him what is the great new discovery in atomic science. Bohr talked for an hour. His friend was disappointed, saying that Bohr had said exactly the same things in 1913. The point is that Bohr had a general (and extremely unconventional) perception of what atomic theory should be like—long before experiment or theory confirmed his "absurd" ideas. By 1930 these ideas were confirmed. But Bohr did not mention this detail. He kept talking about the original contradictions.

At the beginning of the 20th century, paradoxes became unavoidable. The most famous paradox had to do with the radiation of hot bodies. We have already seen that atoms have energy due to temperature. (The probability of finding a certain energy E is proportional to $e^{-E/kT}$.) Starting from this point, one can show that if a gas has a temperature T, then, on average, an atom will have a kinetic energy of $(3/2)kT$.

If atoms carry an electric charge, they can emit or absorb light. They should act like little antennae. Just as there are statistical laws regulating how much energy atoms carry, so there are statistical laws regulating the radiation of atoms. The way physicists think about these laws is to consider a box with reflecting walls. Radiation in this box can have wavelengths in a variety of directions, but each wave must "fit" in the box in the sense that the wave amplitude must be zero at the walls of the box. One has to count the different ways in which oscillations of the wave field may occur. Actually, one can show that if one considers boxes of different volumes, then in twice the volume, there are twice as many possible oscillations.

In each way that oscillations take place, one can show that the average energy content should be kT. How much total energy should we have? We can have shorter and shorter wavelengths. This is terrible! There are an infinity of possible wavelengths and the total energy should be infinity times kT!

Should we really believe that the energy is kT for each wavelength? If we make the box very hot and put a hole in it then we can observe the radiation that comes out. It was observed that the energy is actually kT if the wavelength is sufficiently long.

The sun radiates in a similar manner up to a wavelength in the infrared range which is beyond visibility. It emits most radiation in the yellow part of the spectrum. But, in the blue and ultraviolet parts of the spectrum, there is less and less radiation. The emission of an infinity of radiation does not occur. Why?

Instead of solving this problem immediately, I will present another paradox. Consider a helium atom. How much energy should it have? It can be argued that helium should have the energy $(3/2)kT$, $kT/2$ in the x direction, $kT/2$ in the y direction, and $kT/2$ in the z direction. This follows from statistical mechanics and it is precisely correct.

On the basis of evidence (which I shall not discuss) we know that a helium atom has a nucleus and two electrons. Each of these should have an energy equal to $(3/2)kT$. So, the total energy should be three times $(3/2)kT$. But, it isn't; at temperatures we usually encounter, the electrons are tightly bound to the nucleus and acquire no energy.

Consider a nitrogen molecule which is made of two nitrogen atoms. (We forget about electrons, since we had no luck with them in the previous example.) Each atom should have an energy $(3/2)kT$ so the total energy should be $(6/2)kT$. The observed energy is $(5/2)kT$.*

After making this last observation, the first great American theorist, Willard Gibbs, wrote in his *Introduction to Statistical Mechanics,* "Difficulties of this kind have convinced the author that he builds on very insecure foundation who wants to discuss the structure of matter. I, therefore, will not discuss physics, but restrict myself to developing a mathematical subject, because in mathematics there can be no other mistake than a disagreement between assumptions and conclusions and this, with proper care, on the whole, one may hope to avoid." This quote shows how surprised, indeed how scared, Gibbs was to learn that nitrogen has only "5 degrees of freedom" instead of "6 degrees of freedom."

The paradoxes can be summarized in one phrase: too few degrees of freedom. Too few in molecules, too few in atoms, too few in vacuum itself (the missing energy in radiation). Statistical mechanics which served so well up to a point, now ceases to function.

All these difficulties are solved by one formula. It is precisely the simplest kind of formula, but it makes very little sense. [It is my contention that a formula, to say anything at all must concern at

---

* WT: This argument is too simple.

  ET: (reluctantly) I will elaborate. The nitrogen molecule should have three ways to move as a whole (in x, y, and z directions). It has two ways in which to rotate (around the two axes) perpendicular to the line joining the two N atoms. Energies connected with these modes of motion account for the actual thermal energy of the molecule. But there is a sixth possible motion, vibration. Under normal conditions, this energy is not present.

least three quantities, since a formula concerning only two quantities (i.e., $a = b$) is only an identity.] The formula we are interested in is of the minimal kind. It is $E = \hbar\omega$, where E is the energy, ω is 2π times the frequency (for the moment I will not tell you the frequency of what) and $\hbar$ is a constant, Planck's constant.*

The formula was introduced by Max Planck in 1900 to explain the actual radiation emitted in conditions of equilibrium (to which solar light is a good approximation). The meaning of the formula is that light is quantized. "Quantized" means that light comes in finite units. You can have energy equal to $\hbar\omega$ or to 2 times $\hbar\omega$ or any integer multiple of $\hbar\omega$, but you can't have energy equal to $(1/2)\hbar\omega$. An analogous situation is money. You can have one cent or two cents or 1,000,000 cents, but have you ever seen 1/2 cent?†

This equation $\hbar\omega = E$ is valid in every case, but how does it help? In the case of the box with reflecting walls, we are dealing with light. Light has a frequency. We can use the old type of statistical mechanics only as long as kT, that is, the thermal energy that is available is greater than $\hbar\omega$. Because of the Boltzmann factor, when $kT > E = \hbar\omega$, the term $e^{E/kT}$ for one quantum is close to one and, for increasing numbers of quanta, will vary in a smooth way. The total energy for a given ω can be shown to equal kT. Infrared light is similar to that of the millionaire who has so many pennies that he isn't aware that money comes in indivisible units.

The formula helps in the case of the helium atom also. The electrons want to move around the nucleus with such a high frequency that they can accept energy only in quite big amounts. If kT

---

* WT: Actually, $\hbar = h/2\pi$, and h is Planck's constant and ω is $2\pi\nu$, where ν is the number of waves. Then, by twisting everything about, we have the most impressive formula $\hbar\omega = h\nu$. This confusion seems most unnecessary, but ET claims it is stated in every physics text and since, as everyone knows, ET is a conformist, we have included it.

† WT: I have seen a 1/2-cent postage stamp.

ET: I know of examples in which an energy $\hbar\omega/2$ is discussed. This footnote has the sole purpose to confuse the reader.

is less than the unit $\hbar\omega$, then the electrons don't participate in the thermal motion.

A similar situation holds for the nitrogen molecule. The molecule can accept any amount of energy in its three motions: up, forward, and sideways. These translational motions have no frequency. One may set $\omega = 0$, since no length of time is sufficient to bring the molecule back to the place where it has been. The two rotations have small frequencies and they can accept thermal energies. But in the case of the oscillation, ω is high enough so that $\hbar\omega \gg kT$. The oscillations don't get any energy.

The idea of quantizing light was used by Planck to explain the spectrum of hot bodies. Einstein used the idea to find the heat content of a solid body in which all kinds of oscillations can take place. Quantization, as Einstein noticed, can also be observed directly in the photoelectric effect. Light that falls on a body can tear out electrons. Thus, an electric current is produced. The formula $E = \hbar\omega$ explains the energy electrons receive if light of a certain frequency falls on them. Of course, one must be careful because the apparent energy of the electron is less than the energy originally received. Some of the energy is used to tear the electron out of the solid. More energy is usually frittered away while the electron moves through the solid. Nevertheless, the photoelectric effect showed that light is quantized and indeed demonstrated this fact in a more explicit manner than the spectra explained by Planck.

Why should energy and frequency be related? This puzzle did not stand alone. Another puzzle was the stability of the atom. Chemists know that all hydrogen atoms are alike (with a few exceptions), all helium atoms are alike, and so forth. This was not particularly surprising as long as atoms were considered indivisible. But by 1913, it became clear that atoms are composite.

To study the atom, the English physicist Rutherford shot electrically charged particles through thin foils. The particles went through the material as if nothing were there. A few of the fast charged particles (which were actually α particles produced in the radioactive decay of heavy elements) were sharply deflected. Rutherford suc-

ceeded to explain his results in a quantitative way by a simple model
of the atom.

In the Rutherford model, the atom consists of a heavy nucleus
whose radius is less than one ten-thousandth the radius of the atom.
The nucleus carries a positive charge which is a multiple of the charge
of the electron. In a neutral atom, this charge is cancelled by an
appropriate number of negatively charged electrons which are distributed around the nucleus within the much bigger volume of the
atom. (Rutherford believed that they are orbiting.)

The bombarding α particles are themselves nuclei of the helium
atom (with a charge of 2 units). The α particles carry so much energy
that they cannot be deflected by the light electrons (whose weight is
$1/1840$ of the hydrogen nucleus and even less compared to the four
times heavier helium nucleus). Only when an α particle comes quite
close to an atomic nucleus will it be deflected. From the distribution
of the deflection angles that he found, Rutherford deduced that Coulomb's Law, $F = e_1e_2/r^2$, is valid down to a distance less than
$1/10,000$ of the radius of the atom as a whole (which is about an
Angstrom unit or 10^{-8} cm). From this, Rutherford conjectured that
the hydrogen atom looked like an electron rotating about a proton
with the ratio of the masses 1 to 1840.* The electron and proton, of
course, carry charges which are equal but opposite.

This simple model raised a not-so-simple question. The electron
may move on a circle around the proton. It is charged and so it
should behave like a little antenna and radiate electromagnetic waves,
that is, light. Thus, it should lose energy and move closer to the
nucleus. The closer it moves, the more it is accelerated and the more
it radiates. The more it radiates, the closer it moves to the proton.
It falls into the nucleus much faster than you can say, "the paradoxes

---

* WT: How did he know the mass of the electron?
   ET: Millikan measured the charge; the charge-to-mass ratio had been measured before
   when electrons of known energy were deflected in magnetic fields.

of atomic theory." In fact, it falls into the nucleus in 10^{-9} seconds (called a nanosecond).

Why is Rutherford in trouble, but not Newton? Shouldn't the earth also radiate and, so, fall into the sun? The earth carries no electric charge and so it will not radiate electric waves, but it does radiate gravitational waves. This gravitational radiation is a consequence of Einstein's theory of curved space. Fortunately, the gravitational waves are very weak and so it will take a very long time for the earth to fall into the sun. It will happen in a time equal to one billion times T, where T is the age of the universe. So don't worry. Before that happens, many other events will occur.

So we come back to the question: Why are atoms stable? Again, I'll ask more questions before I answer the last one.* Atoms absorb and emit light. This light is characteristic of the kind of atom. If I throw sodium into a flame, the flame will be yellow. But if I throw lithium into a flame, red light is emitted. This is spectroscopy. I can identify an atom by the precise wavelength of the light it emits or absorbs. In the spectrum of the sun there are "Fraunhofer lines," actually black lines.[†] The last adventure of light before it leaves the sun is to run the risk of being absorbed by sodium (or some other element). So, we get black lines. By looking at the spectrum of the sun we can find out the composition of the sun, which turns out to be similar to the composition of the earth. Similarly, we can look at the spectrum of other stars to find out their composition.

Each atom is "tuned." It has definite frequencies. Atoms have many frequencies. These frequencies are related in simple ways, say, if ω_1, ω_2, ω_3, and ω_4 are characteristic frequencies, it is often the case that $\omega_1 + \omega_2 = \omega_3 + \omega_4$. The tuning of atoms is different, however,

---

* WT: I believe that you are guilty of changing the topic whenever you do not know the answer to a question.

ET: Two paradoxes are better than one; they may even suggest a solution.

[†] One famous explanation of the Fraunhofer lines (given by a harassed student in an examination) was: "The Fraunhofer lines are black lines in the spectrum of the sun emitted by elements absent in the sun." The explanation we shall give is less concise and lacks the literary brilliance of the student's answer. However, it is correct.

from the tuning of a musical instrument. If a string has a frequency of ω, it will also have frequencies 2ω, 3ω, and so forth, which are called "overtones." Multiples of frequency are usually not found in "atomic tuning."

What we just said adds to the difficulties. All hydrogen atoms have the same frequencies. Why don't we find hydrogen atoms with electrons in any orbit? Each orbit should have a different frequency. Between all the hydrogen atoms, all frequencies should be emitted. Why are the atoms stable? Why do they have sharp lines? Why are there peculiar relations between the frequencies? Why? Why? Why?

Bohr tackled these problems twelve years after Planck. His explanations were characteristic; they were crazy. People did listen, probably because it was clear that no easy answers were going to come. Bohr did not explain Planck's formula ($E = \hbar\omega$). He said it is a strange atomic law which holds. It is simple, so simple that it could not be explained further by anything more simple. So, it should not be explained at all. It should not be the end point of atomic theory. It should be the starting point.

Don't let this apparent simplicity go to your head. Bohr certainly didn't become overconfident. In 1933, there was a congress of philosophers in Copenhagen. By that time, the formula $E = \hbar\omega$ was universally accepted. The consequences were worked out and became quantum mechanics, the kind of mechanics that recognizes quanta. The congress was quite a cordial affair. Everyone accepted what Bohr said. The morning after the congress, Bohr appeared with a scientific hangover. Nobody understood why. Finally, he explained, "Anyone who can say $E = \hbar\omega$ without becoming dizzy doesn't know what he is talking about."

The second thing Bohr said was that while atomic laws should not follow from the laws of macroscopic physics, they should correspond to those laws. One should be able to guess atomic laws by the way things act when much more energy is available than $\hbar\omega$. This is the "correspondence principle." It is important in the history of physics and in understanding the philosophy of physics.

The last point Bohr made was to postulate not only $E = \hbar\omega$, but also the stability of atoms. In fact, he postulated the stability of many atomic states.

Now it really seems that Bohr has gone too far. His theory is full of unsupported assumptions which, moreover, seem absurd. Can one expect anything to follow that is logical, or at least consistent?

Bohr said that he had no choice. The model of the hydrogen atom is so simple that no amount of mathematical trickery could help. All that one can hope for is to find some pattern in all the contradictions (a difficult job since the scientific mind is not particularly compatible with contradictions). Yet, the conclusions drawn are straightforward enough.

Since light has energy $\hbar\omega$, one must say that the energy difference between two states is $\hbar\omega$. Then, if an electron jumps from one state to another, it will emit or absorb light with energy $\hbar\omega$.

The postulate of stable levels explains the fact that we have relations of the kind $\omega_1 + \omega_2 = \omega_3 + \omega_4$. If we multiply both sides by $\hbar$ we get $\hbar\omega_1 + \hbar\omega_2 = \hbar\omega_3 + \hbar\omega_4$ or $E_1 + E_2 = E_3 + E_4$, where E_1, E_2, E_3, and E_4 are energy differences between appropriate levels. Let us draw in Figure 1 four levels, where the distance from the bottom represents energy. We see at a glance that the relations $E_1 + E_2 = E_3 + E_4$ makes sense.

But what is ω? Is it the frequency of the electron in the lower state or the frequency of the higher state? We run into our first real problem. Bohr has given us a classical model of the atom to relate to the data we want to explain. But, he does not tell us how to relate the frequency of a *transition* to the energy of a *state*.

In highly excited states, the frequencies of neighboring states are similar. The result is that the levels between which transitions occur are similar in energy and, hopefully, similar in their other properties. Thus their frequency of the transition is nearly equal to the frequency in the initial state and nearly equal to the final state: the two are quite similar.

The paradoxes are strong and obvious. The postulates of stable states and the relation between energy and frequency seem unnatural (though simple). The correspondence principle may give one a little

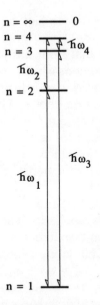

Figure 1. This sketch of the energy levels for the hydrogen atom illustrates how $\omega_1 + \omega_2 = \omega_3 + \omega_4$. At the top, the energy is 0; the electron escapes—the atom is ionized.

reassurance that one will not completely break away from the "real," macroscopic world of our senses, but it seems at best a vague guideline to the way in which the laws of atomic science should be established. But, the last point works better than one might expect. It is best to consider the simplest example. This is the harmonic oscillation.

I mentioned that Galileo became interested in physics when he saw a chandelier swing in church and he noticed that it swung with the same frequency no matter what the amplitude. In fact, it is one of the elementary laws of classical mechanics: if the restoring force is proportional to the displacement, then the frequency is independent of the amplitude.

In a hydrogen atom, the electron moves around the nucleus so the force should depend on the distance r as $1/r^2$. The motion is quite different from a harmonic oscillation. But, consider the di-

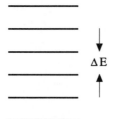

Figure 2. This sketch illustrates the fact that the energy levels of the harmonic oscillator are equally spaced, so the energy difference between levels remains constant.

atomic molecule (diatomic meaning two atoms in each molecule) of nitrogen, N_2. The two nitrogen atoms want to be at a certain distance from each other. If I push them together, they will try to get back to their original position and the more they are compressed, the harder they will push back. Here we do have a force which is (in first approximation) proportional to the displacement. So, we have a harmonic oscillator. The frequency of a harmonic oscillator should be independent of the energy, so the levels should be equidistant and should look like the levels shown in Figure 2. The energy difference is the same for each pair of neighboring levels because the frequency of the light which is absorbed or emitted will remain the same in the higher states of energy.*

Unfortunately, this doesn't completely correspond to experimental data. It turns out that the more strongly a diatomic molecule oscillates, the smaller the frequency becomes (because the restoring forces do not continue to increase sufficiently with the displacement). But this happens very slowly. So, we modify the level scheme to look like Figure 3. (Actually, the figure is exaggerated. Usually, the distance between levels decreases more slowly than has been shown.)

If we consider rotation, we know that the frequency of the ro-

---

* Incidentally, when we have a transition between second neighbors in a harmonic oscillation, the frequency is 2ω which "corresponds" to an overtone in classical physics.

Figure 3. To correspond with experimental data, the energy levels for the diatomic molecule has to be slightly modified so that the levels slowly become closer as we approach the top where the energy = 0, and the molecule comes apart—that is, it dissociates.

tation depends on the energy. But, in rotation, the energy is proportional to v^2, where v is the velocity. The higher the energy, the higher is the velocity and so the frequency is higher. The frequency, and also the distance between levels, should increase with the square root of the energy and we get a level scheme as shown in Figure 4. If we label the levels with integers l, as quantum numbers, then the distance between levels should be proportional to l. The frequency for each level should be proportional to the distance which is proportional to l. The energy should be proportional to the sum of the frequencies, which is proportional to l^2 (see Question 1).

Using this sort of argument, one can get the level scheme for hydrogen. A high school teacher had fitted the spectrum of the hydrogen atom by the formula $R(1/n_1^2 - 1/n_2^2)$, where R is called the Rydberg constant and n_1 and n_2 are two integers. This agreed with Bohr's ideas and gives for the energy level of hydrogen $E = \text{const}/n^2$. The frequencies for transitions between neighboring levels would then be $\text{const}(1/n^2 - 1/(n + 1)^2) \approx 2/n^3$. Therefore, frequency $\approx (\sqrt{E})^3$. Does this agree with the correspondence principle?

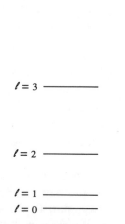

Figure 4.   The completed theory of quantum mechanics gave energy levels somewhat different from the squares of the quantum numbers which are used in this sketch.

The potential energy is proportional to $1/r$. The equation for the centrifugal force $mv^2/r = e^2/r^2$ shows that the kinetic energy, $mv^2/2$, is also proportional to $1/r$. For the frequency $\omega = v/r$ we get $m\omega^2 = e^2/r^3$, so ω changes with the third power of $\sqrt{E}$. Even the Rydberg constant can be accurately calculated from the known values of e, m, and $\hbar$.*

People developed Bohr's ideas. They had the correspondence principle and data. The data, the spectra of atoms and molecules, the laws of chemistry, constituted an overwhelming body of material. The correspondence principle seemed dreadful, it was shaky and ill-

---

* WT: Is this the way Bohr did it?
  ET: Yes.
  WT: Did he tell you that?
  ET: No. But I am convinced he cheated. The Balmer formula (with similar formulas applying to the spectra of alkali atoms) gave him the idea of energy levels. Then he tried out whether or not the connection between energy and frequency checked. Even so, it was a fantastic leap of imagination. But the most remarkable fact was that Bohr knew where to stop, what *not* to explain.

defined. When, after a dozen years, a consistent mathematical solution was found, most people turned their backs on the correspondence principle. The years 1914–1925, from the publication of Bohr's
most unusual paper to Heisenberg's invention of a new kind of mechanics, called quantum mechanics, was a unique period in the history of physics. I started to study with Heisenberg two years after
the end of this period, and I sensed the rapidly fading memory of
those turbulent years. The influence of Niels Bohr, reinforced by
experience assuming an increasingly regular shape—under conditions
which were in fact self-contradictory—were both unacceptable and
incredibly stimulating.

A monument to the psychological conditions of those times is
a paper by Bohr, Kramers, and Slater. They postulated that energy
is not conserved, except in a statistical sense. The experiments of
Compton, on the scattering of x rays on electrons almost immediately
disproved this paper, by demonstrating that energy and momentum
were conserved in the individual quantitative process of scattering
of x rays. In the years 1929–1935, I listened to Bohr's lectures and
mutterings for many hours. He never once breathed a word about
that paper.

I feel that the correspondence principle should not be forgotten.
It is historically important. Secondly, the mathematical principles
involved in quantum mechanics are so complicated that you must
work a year (or even years) to get a solution to a problem. Even
then, you are not sure whether the answer is correct. You can use
the correspondence principle as a check.

Quantum mechanics is complicated. How can I understand it?
What do I mean when I say "understand"? I understand the geog-
raphy of a city if I am familiar with its main streets and where they
are situated in relation to each other. Then, if I want to find a particular place, even though I've never been there, I can find my way
with just a few instructions.

I think about understanding a theory in physics in the same
way. I understand it if I know the main ideas and how they fit together. Then I can get to a "new place" in the theory with the help
of a few "connections" with the main ideas. I can understand classical
physics in this sense of "understanding." It is not surprising to me

that $F = ma$, because I have experience with force and acceleration every time I drive a car. It is almost impossible to have this kind of "understanding" in quantum mechanics because I do not have experiences in the "microworld" of the atoms. To gain this kind of understanding in quantum mechanics, I need the correspondence principle which gives us the needed footholds and handholds to scramble up the precipitous walls of the new theory (unless you want to go by the brand-new funicular, i.e., the mathematical formulism*).

QUESTIONS

1. Show that the sum of the numbers from 1 to n is (nearly) proportional to n^2. What does this mathematical exercise have to do with the energy of a rotating diatomic molecule?

2. Consider the hydrogen atom. It is an electron revolving about the nucleus. Find the time of revolution as a function of r, the radius of the orbit of the electron. Then, write the frequency as a function of the energy. In this way, show that the levels of hydrogen can be written as $-k/n^2$, where k is a constant and n is an integer.

3. Consider the rotational states n and $n + 1$ of a diatomic molecule. What is the difference in angular momentum values?

4. Consider circular orbits of an electron in a hydrogen atom. What is the difference in the angular momentum values for the states n and $n + 1$?

---

* The usual spelling is "formalism." But the spelling used here expresses a justified behavior in a more expressive fashion.

Chapter 10

WAVE–PARTICLE DUALISM

*In which the structure of matter is explained and
chemistry is unified with physics.*

About a dozen years after Bohr discussed the correspondence principle, a French student in Paris, Louis de Broglie, handed in his
doctoral thesis. The thesis looked like nonsense; it said that electrons
are really waves. It was not surprising that the first inclination of the
faculty was to reject the thesis, especially since the faculty of the
Sorbonne was somewhat conservative. The situation was, however,
not so simple. De Broglie came from a noble family; he was a prince.
Furthermore, his father was a powerful politician who had been prime
minister. After the French Revolution, princes were a little less respected, but politicians much more so.

Fortunately Einstein happened to be in town at the time and
so the thesis was shown to him. He was enthusiastic about the idea
and sent copies to all his friends. On Einstein's say so, de Broglie

received his degree. This was very helpful, because de Broglie later received the Nobel Prize. It would have been embarrassing had it been awarded on a thesis that had been considered unacceptable by the Sorbonne. Strangely enough, de Broglie never did any really outstanding work again.

The idea that electrons had something to do with waves was worth not just one Nobel Prize but, in fact, two. You will remember that von Laue illuminated a crystal with x rays, which everyone accepted as waves. The x rays were deflected by the crystal and gave an interference pattern. Instead of x rays, electrons could be used to bombard a crystal and a similar interference pattern was observed. This was experimental proof that electrons were waves, and this proof (by Clinton Davisson and Sir George Thomson) actually received a second Nobel Prize in 1937.

In his thesis, de Broglie's idea was to show that electrons are waves and not particles. Actually, the development of quantum mechanics, which is our present complete description of the atomic world, maintained that electrons—like protons and all other subatomic particles—must be considered as entities that have both particle properties and wave properties. This was a radical break with the simple and old tradition: namely, that one fact has but one explanation. Yet this duality has proved extremely fruitful and I consider it just as final as the earlier great revolution—which asserted that the earth moves around the sun rather than the other way around.

Of course, this duality would be absurd if the behavior of waves and of particles were radically different. The remarkable fact, however, is that waves and particles are surprisingly similar in their behavior. To understand the dualism, these similarities have to be emphasized, and therefore, we shall proceed to do so in detail.

De Broglie himself started from a similar behavior of particles and waves in relativity. Free particles are characterized by their energy and momentum. The four quantities—the energy and the three components of momentum: the momentum in the x direction, the momentum in the y direction and the momentum in the z direction—can appear to be different for different observers. In fact, earlier we have seen that they are the components of a four-dimensional vector. (Recall that in the relativistic transformation, energy behaves like time and the three components of the momentum behave like the three space coordinates: x, y, and z.)

In the wave description, we also encounter the components of a four-dimensional vector: the frequency, which is the number of vibrations per unit time, and the wave number, which is the number of maxima encountered when we proceed a certain distance in the x, y, and z directions.

In the particle picture, we talk of the four components: E, p_x, p_y, and p_z. In the wave picture, we talk of ω, k_x, k_y, and k_z. ω is the frequency, or more precisely, the number of vibrations occurring in 2π seconds.* k_x, called the x component of the wave number, is the number encountered if you proceed 2π centimeters along the x axis. It is easy to see that if you simply rotate your coordinate system, k_x, k_y, and k_z transform as the components of a three-dimensional vector.

Having introduced these definitions, de Broglie could now write

$$E = \hbar\omega$$

$$p_x = \hbar k_x$$

$$p_y = \hbar k_y$$

$$p_z = \hbar k_z$$

In this simple way, the *momentum* of the electron, considered as a particle, can now play a role which has a counterpart in the wave description. Peculiarly enough, the story about the *velocity* of the electron is more complicated. Staying with the relativistic de-

---

* WT: This introduction of 2π seems to be just the physicists' way of complicating things, and it means that the wave number also has to have a 2π in it so that the equations work out correctly.

ET: You can also explain the introduction of 2π by noting that you can speak of "radians per second" when you talk about frequencies, and *that* explanation does not denigrate physicists.

scription, the energy of an electron at rest is mc^2, where m is the mass of the electron. Since the electron is at rest, its momentum is zero, and this must mean that $k_x = k_y = k_z = 0$. But the velocity of a wave, which is the wavelength over the period, shows a great peculiarity for electrons at rest. The period, the reciprocal to the frequency, will be finite. But the wavelength, the reciprocal to the wave number, must be infinite. Thus, an electron at rest should have an infinite velocity as a wave!

At this point, de Broglie made an obvious, simple, but remarkable contribution. He utilized the distinction between the phase velocity of waves and the more complex, but more important, group velocity.

The phase velocity is what people at first sight associate with the "wave velocity." The phase velocity is the speed with which individual wave maxima propagate. This can be made clear if we write, for a plane wave propagating in the x direction, $\cos(kx - \omega t)$. If we pick a maximum of the wave amplitude, then there the cosine must be one and the phase zero. Here, then, $kx - \omega t = 0$, or $x = \omega t / k$. Thus the phase velocity is $v_{\text{ph}} = \omega / k$.

For light in vacuum, the velocity is always equal to c and, knowing this velocity, we have $\omega = ck$ and nothing more need be said. But with electrons the situation is vastly different. In de Broglie's relativistic discussion the energy of an electron at rest is mc^2 and its frequency is $\omega = mc^2 / \hbar$. Being at rest, its momentum, and wave number are zero ($k = 0$) and so the phase velocity ω / k turns out to be infinite, as stated above.

Here comes the essential contribution: If a wave number is to represent or correspond to a particle, the wave process must be localized. This can be done by a bunching up of wave crests or forming a wave group as shown in Figure 1. One can show that such a group can be formed by adding up or superposing plane waves of somewhat different* wave numbers so that in the center of the group all am-

---

* If the wave numbers are only slightly different, then the reinforcement will occur in a broader region instead of just the one spot where we want it to happen.

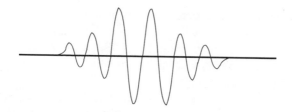

Figure 1. This wave group can be formed by adding together several plane waves of differing wave numbers.

plitudes reinforce each other while outside the group they cancel. De Broglie was not clear about the ultimate consequences of his ideas: the square (of the absolute values) of his waves were to be identified with the probability of finding an electron. So the velocity of the electron should be set equal to the velocity with which the group maximum is moving. That velocity is the group velocity. (That point de Broglie saw very clearly.)

We can determine the group velocity by noting that the crest of the group is the point where all the phase differences of the component waves remain zero. To be quantitative, we start with the phase difference between two waves with ω_1, k_1 and ω_2, k_2 which is $(k_1 - k_2)x - (\omega_1 - \omega_2)t$. This quantity must remain zero at the crest, although individual wave maxima may overtake (or be left behind) the crest.

The velocity of the point where the phase difference remains zero is clearly given by $(\omega_1 - \omega_2)/(k_1 - k_2)$. Because the difference between the two wave numbers has to be small, the ratio of the two differences is an approximation to a derivative. When we have lots of waves being added together, the group velocity is thus $v_g = d\omega/dk$. Notice that in vacuum, for light, the velocity doesn't depend on ω and $d\omega/dk = \omega/k = c$; the group velocity and the phase velocity are the same and are both equal to c.

In nonrelativistic physics, for particles, $E = p^2/2m$, and $v_{ph} = \omega/k = E/p = p/2m = v/2$, which is one half of the velocity of the particle. But $v_g = d\omega/dk = dE/dp = p/m = v$, which is the correct value for the particle velocity. In the relativistic case,* $E^2 = p^2c^2 + m^2c^4$, so $v_{ph} = \omega/k = E/p$, which is infinite for $p = 0$ and is c at high energies (as it should be). But $v_g = d\omega/dk = dE/dp = pc^2/E$, which is the correct particle velocity—zero at $p = 0$ and approaching c at high energies.

Where does this leave us? De Broglie says that electrons, which we always knew as particles, are waves. But Einstein already proposed that light, which we know as a wave process, behaves in some cases as though made up of particles or quanta. We seem to be getting closer. Two related difficulties are better than one.[†]

Actually, the debate, whether light is a wave or whether it is made of particles, started in the time of Newton and the debate has a direct bearing on modern physics. It is a well known fact that a light beam, when it hits water, changes it direction. Willebrod Snell discovered this law of refraction in 1607. Newton explained this phenomenon by saying the light is made up of particles which are attracted into the water. Huygens, a Dutch contemporary of Newton, explained the refraction of light by saying that light is a wave that propagates with different velocities in the two media. Both theories could explain the observation. Neither could disprove the other's argument. The discussion was a draw. Remarkably enough the two arguments are essentially the same; the difference is hardly more than a difference in language. This circumstance was not appreciated until, a couple of centuries later, the wave–particle dualism came into the focus of discussion.

First, we consider the empirical law of refraction, that light bends as it travels from air into water. We call the angle between the light

---

* See the Questions section at the end of this chapter.

† One may guess at Einstein's feelings when his colleagues in Paris presented him with the "crazy" wave of young de Broglie. In 1905, when Einstein made his great contributions to the determination of the size of the atom (Brownian motion), when he talked about light quanta, and when he invented relativity, he was outside the academic community. It would be nice to say that Einstein always recognized the truth when he met it. Unfortunately, Einstein refused—in the end—to accept quantum mechanics, the field of physics that grew out of de Broglie's thesis. Physics is, hopefully, simple. Physicists are not.

beam in air and the vertical to the surface α_i for "α incident" and we call the angle the light in water makes with the vertical α_r for "α refracted." The law of refraction (called Snell's Law) states that the ratio of $\sin\alpha_i/\sin\alpha_r$ is always the same, no matter what angle of incidence is chosen.

Newton's explanation of refraction went like this: Light is made of particles, "corpuscles," which have momentum. There is a force exerted by the water surface that pulls the corpuscles into the water. The momentum vector has two components, one parallel to the water surface, $p_{\parallel}$, and the other perpendicular to the water surface, $p_{\perp}$. It is natural to assume that the force is vertical to the surface so the horizontal component $p_{\parallel}$ will not be affected. The vertical component, however, will change—it will increase as the force acts on the corpuscles. The kinetic energy of the incident particle is, of course, independent of its direction of incidence. The change in the kinetic energy is equal to the negative of the change in potential energy as it enters the water. So, it must be independent of the angle of approach. Thus the ratio of the kinetic energies and also the ratio of momentum values is the same for all angles. That means that p_i/p_r will remain the same but the ratio of the parallel components is one; $(p_{\parallel})_i/(p_{\parallel})_r = 1$. We see from Figure 2 that $(p_{\parallel})_i = (\sin\alpha_i)p_i$ and $(p_{\parallel})_r = (\sin\alpha_r)p_r$. It follows that $\sin\alpha_i/\sin\alpha_r$ is the same for all angles of incidence and can be called, according to Snell, the refractive index, n. So the law of refraction follows from the particle theory.

Huygens believed that light was a wave. As it passes from air into water, its frequency cannot change—the number of crests passing some arbitrary observation point per unit time remains the same. But the velocity of propagation* is different in the two media and so the wavelength—and therefore the wave number—will be different. But the change in the wave number from the one medium to the other will be such that the ratio of the wave numbers will be the same, no matter what the angle of incidence. We notice that in the

---

* That is the phase velocity. According to our earlier discussion, this is *not* the particle velocity.

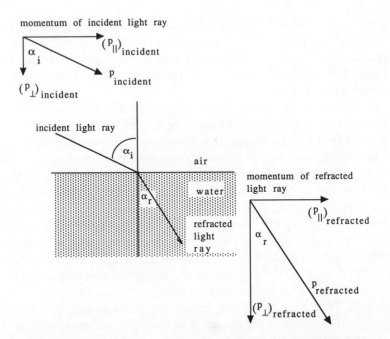

Figure 2.   The momentum p of the ray of light changes as it passes from the air into the water, but component $p_{\parallel}$ along the surface remains unchanged.

particle theory the ratio of momentum values remains the same, while in the wave theory the same holds for the ratio of the wave number values. In the former case, the original argument depended on considerations of energy, in the latter case on considerations of frequency.

It is easy to understand the situation corresponding to the statement in the particle theory that the component of the momentum parallel to the surface must be the same in air and in water. If we consider the wave on the air–water boundary, the requirement of continuity means that the number of crests encountered on either side of the boundary must be the same. Following the dashed line above and below the interface between air and water in Figure 3, we encounter the same number of crests in air and in water, and there-

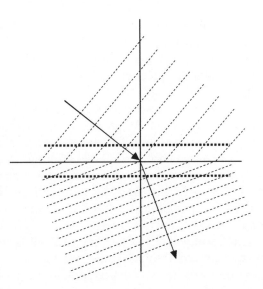

Figure 3.   The $k_\parallel$ components of the incident and refracted waves remain unchanged, as can be seen by examination of the wave crests above and below the interface.

fore the component of the **k** vector parallel to the surface remains the same.

Since the ratio of the **k** vectors is fixed and the horizontal components are the same, Snell's law follows in wave theory just as it did follow in particle theory. Newton's argument that $p_r/p_i = n$, is a constant, becomes $k_r/k_i = n$, the same constant.

The particle explanation and the wave explanation do not merely lead to the same result (Snell's law). The two explanations are the same, except for the vocabulary. One is in plain English (Newton energy momentum), the other is in straightforward Dutch (Huygens's frequency wave number).*

So far we have only used the statements that energy and fre-

---

* WT: Actually, didn't both of them write in Latin, n'est-ce pas?
  ET: Yes, but they had to explain their arguments in their native tongues, probably.

quency (and also momentum and wave number) correspond. With the advent of quantum theory, energy and frequency became proportional. Momentum and wave number are also proportional. Moreover, the constant of proportionality is the same. This will become clear in the next example.

In the example that we have discussed so far, particles, or waves, moved in a fixed surrounding. What happens if there is movement in the surrounding, for instance if particles, or waves, are reflected from a moving (material) wall? We know what the result will be: If you throw a ball at a stationary wall and it makes a perfectly elastic collision, it bounces—reflects—off the wall with exactly the same velocity with which it approaches the wall. If the wall is moving toward the ball (say the wall is a batter's bat and you are pitching a baseball), the relative velocity of the bat and ball is still the same. But to you, the pitcher, the ball comes off the bat with a higher velocity than you threw it. Now let us look at this in a little more detail.

Suppose a particle is moving towards a wall and the wall is moving toward the particle as in Figure 4. For sake of simplicity, we assume that the particle moves perpendicularly toward the wall.

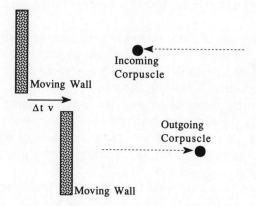

Figure 4.   If a particle strikes a wall that is moving toward it, the particle gains energy from the collision such that the relative velocity of the particle and wall is unchanged.

The result of the collision will be that after it hits the wall the particle will rebound with a larger velocity and a larger kinetic energy. Let the wall have a big mass M and the velocity v_w. Then its change in momentum is the same magnitude but of opposite sign as that of the particle: $\Delta p_w = |p_i| + |p_o|$, where p_i is the incoming particle momentum and p_o is the outgoing particle momentum. The change of the kinetic energy of the wall is the same as that of the particle: $|\Delta E| = (\Delta p_w)v_w$, where p_w is the momentum of the wall and v_w is the wall's velocity—assumed unchanged because of the huge mass of the wall, relative to the particle. The kinetic energy of the particle must change by the same amount as the kinetic energy of the wall as $E_o - E_i = (p_o + p_i)v_w$,* where E_o and E_i are the outgoing and incoming energies. This is the essential equation.

$$E_o - E_i = (p_i^2 - p_o^2)/2m = (p_i + p_o)(p_i - p_o)/2m = (p_i + p_o)v_w$$

and therefore $(p_i - p_o)/2m = (v_i - v_o)/2 = v_w$, where the v_i and v_o are the absolute values of the incoming and outgoing particle velocities. It follows that $(v_i - v_o) = 2v_w$ or $v_o - v_w = v_w - v_i$. Considering the different directions of v_o and v_i, the last equation simply means that the relative velocity with respect to the wall has remained the same before and after reflection. We obtained that obvious result in a complicated way, using conservation of momentum and energy, because we now proceed to get the same result in the wave-theory by the same argument.

We now ask, can Huygens explain the situation as well? Instead of a particle moving toward the wall, consider a wave moving toward the wall. We expect that the wave will rebound from the wall with a different frequency and different wavelength. These sorts of changes occur when a train approaches us and we hear the high pitch of its whistle drop suddenly as the train passes us and then recedes from us—the Doppler effect.

---

* In the rest of this derivation we write p_o and p_i for $|p_o|$ and $|p_i|$.

To find how the frequency incident on the wall ω_i is related to the outgoing frequency ω_o and how these in turn are connected with incident and outgoing wave numbers k_i and k_o, we count the wave crests that pass per second through a plane that is some distance from the wall, a plane toward which the wall is moving. At first you might think that the same number of crests must be moving toward the wall as away from it because no crest could have been lost. That would give $\omega_i = \omega_o$. But actually in one second the wall moved the distance v_w toward our plane. Therefore after the passage of one second there are fewer crests of the incoming wave left (in the amount $v_w k_i$) and also in the outgoing wave there are fewer crests left (in the amount $v_w k_o$). Altogether there are $v_w(k_i + k_o)$ crests missing and these must have escaped through our plane. The number of crests leaving, ω_o, must be greater by $v_w(k_i + k_o)$ than the number incoming, ω_i. We have, therefore, the relation $v_w(k_i + k_o) = (\omega_o - \omega_i)$.* This looks just as the result $v_w(p_o + p_i) = E_o - E_i$, which we obtained from the elastic reflection of a particle from the wall. Indeed, if we multiply the equation connecting the ω and k values by $\hbar$ and remember that $E_o = \hbar\omega_o$, $E_i = \hbar\omega_i$, $p_o = \hbar k_o$, and $p_i = \hbar k_i$, the two equations turn out to be the same. It is enough to know that $v_w(k_i + k_o) = (\omega_o - \omega_i)$, together with the equations connecting ω values and k values, in the same way as E and p are connected, to get the laws of wave reflection from the wall. Wave theory and particle theory again correspond. But to get the correspondence, it is now not enough to say that energy and frequency are connected, as well as momentum and wave number. We must add that energy is proportional to fre-

---

* WT: Don't we have to take the wall advance as actually $2\pi v_w$ seconds if we are to use $v_w k_i$ and $v_w k_o$ crests?

ET: Yes! This is terrible. So we have to wait 2π seconds. And it is just 2π seconds in which the difference between outgoing and incoming crests will be $\omega_0 - \omega_i$.

WT: So you claim your final equation is right. I knew it. This is how you "simplify" all the time. Mathematicians have a function, called a Shlomomorphism (named for a physicist in the clothing of a mathematician who was a teacher of mine) which maps any quantity onto itself multiplied (or divided) by 2π. ET has just committed a Shlomomorphism.

quency, and momentum to wave number, with the same constant of proportionality.

We have produced this result without appeal to relativity,* although one argument was given in Einstein's relativistic English, the other in de Broglie's dual French.

We can now settle the old controversy of wave versus the particle theory of light, like good fence sitters, by saying both sides were right. The same holds for electrons or any other objects.†

The idea that something can be described in two incompatible ways is not new, it is only new in physics. It is an old idea in theology. From my point of view you are a body, but from your point of view you are a soul. Both descriptions are valid. Perhaps the fact of wave–particle duality will make us more tolerant of the mysteries of the body–soul duality. Niels Bohr was convinced that the body–soul duality will eventually be resolved in a manner related to the resolution of the particle–wave duality. He even believed that every problem of real interest contains an element of duality.

One of the people who received de Broglie's thesis from Einstein was Schrödinger. It was Schrödinger who used the wave concept of the electrons to explain the stable states of atoms. Among the many orbits for a particle, classical physics makes no choice. Schrödinger showed that if the electrons in orbit about the nucleus were thought of as wave forms, only some fit. As in Figure 5a, the wave function describing that electron does not fit, but another does as in Figure 5b. (The same situation occurs in classical physics in the case of vibrating strings, only certain wavelengths and frequencies are permitted.)

In order to allow the electron to be in an orbit, you must have an integral number, n, of wavelengths fit into the orbit, $2\pi r / \lambda = kr$

---

* WT: The classical result seems to have taken a lot more explaining than did de Broglie's relativistic approach.

ET: Did I not say, right at the beginning of this book, that relativity makes things simpler?

† WT: But they can't both have been right!

ET: Actually, they were, but more important, neither was wrong.

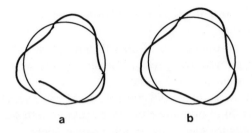

a   b

Figure 5.   An electron can only be in an atomic orbit if its momentum is such that its de Broglie wavelength fits an integral number of times around the orbit. In (a) the wavelength is too large; in (b), three wavelengths fit exactly.

$= n$, which implies $pr = n\hbar$, which is the quantum condition that Bohr derived from the correspondence principle.

We have now four interpretations of waves in terms of particles:

If a wave process is restricted to a part of space, then the particle associated with that wave is to be found in that part of space.

If a wave has a frequency ω, the particle associated with the wave has the energy $\hbar\omega$.

If a wave has the wave number k, the particle associated with it has the momentum $\hbar k$.

If a wave fits onto an orbit within an atom, that orbit is permitted.

This last statement is obviously qualitative. How the other three statements hang together is not clear either, except in special applications which have been discussed. A general, consistent interpretation is needed. This came from the very pure mathematics of Hilbert, the physics of Heisenberg, and from John von Neumann, who put the two together in axiomatic form to show the absence of any contradictions.

What we now have to describe has two difficulties. One is duality, the need for two contradictory approaches. The other is the use of abstract mathematics, the theory of linear differential equations, which is not close to common experience.* The idea we have to introduce is that of a linear operator. In itself, it is not a complicated idea.[†] It is at this moment entirely unclear what it is good for. Yet it turns out that it will result in a method that allows us to extract precise statements about a particle from a wave.

The first concept we have to introduce is an "operator." An operator is a procedure that makes out of a function another function. Our interest is in waves; for instance we may consider $\psi(x, y, z, t)$, where ψ is a function that depends on the position (x, y, z) of a particle at the time t. An example of an operator would be the procedure to replace the function ψ by $1/\psi$.

The second step is to restrict the kind of an operator we want to consider. Our operators should be linear. This means that if the operator acts on the sum of two functions, the result is the same as letting the operator act on the two separate functions and then adding *that* result. Let us designate the result of an operator $\mathcal{O}$ acting on ψ as $\mathcal{O}\psi$.

Is the example given in the previous paragraph a linear operator? No!

Let the procedure, "take the reciprocal," act on ψ. Then $\mathcal{O}\psi = 1/\psi$. Let it act on another wave function ϕ, so that $\mathcal{O}\phi = 1/\phi$. Let now $\mathcal{O}$ act on the sum of the two, $\mathcal{O}(\psi + \phi) = 1/(\psi + \phi)$. This is not the same as the sum $\mathcal{O}\psi + \mathcal{O}\phi = 1/\psi + 1/\phi$. So we had given an example of an operator, but not of a linear operator.

On the other hand, let us take as our procedure the action of differentiation with respect to x, leaving the other variables unchanged: $\mathcal{O}\psi = (\partial/\partial x)\psi$. This is a linear operator because

---

* WT: What did Copernicus say?
  ET: "Mathematics is for the mathematicians."
  WT: Was he right?
  ET: Dualism is democratic. Even mathematicians must be listened to.
[†] WT: You mean as simple as curved space?
  ET: Much more simple.

$$\frac{\partial}{\partial x} (\psi + \phi) = \frac{\partial}{\partial x} \psi + \frac{\partial}{\partial x} \phi.$$

Therefore $\mathcal{O}(\psi + \phi) = \mathcal{O}\psi + \mathcal{O}\phi$.

Finally we need the concepts of eigenfunction and eigenvalue. If, as a result of applying an operator, the function is multiplied by a constant factor F, then that function is an eigenfunction (or "proper" function) of the operator and the factor F is the eigenvalue or proper value. For instance, the operator may be the differentiation $\partial/\partial x$.

Then e^{Fx} is an eigenfunction and F is the eigenvalue, because $(\partial/\partial x)(e^{Fx}) = Fe^{Fx}$.

Now we are ready to make a statement about physics,* though unfortunately not about an entirely reasonable wave function. Our simplest (!) example is $\psi = e^{i(kx-\omega t)}$. The imaginary factor i is, of course, the square root of -1, that is $i = \sqrt{-1}$. What we have written is obviously a wave which propagates in the x direction with the phase velocity ω/k (we mean that ψ remains the same if t is increased by Δt and x by $\Delta x = \Delta t \, \omega/k$). The reason to choose this ψ function as our first example is that it is the eigenfunction of two important operators $(\hbar/i)(\partial/\partial x)$ and $-(\hbar/i)(\partial/\partial t)$. Indeed,

$$\frac{\hbar}{i} \frac{\partial}{\partial x} e^{i(kx-\omega t)} = \hbar k e^{i(kx-\omega t)} = p_x e^{i(kx-\omega t)}$$

and

$$-\frac{\hbar}{i} \frac{\partial}{\partial t} e^{i(kx-\omega t)} = \hbar \omega e^{i(kx-\omega t)} = E e^{i(kx-\omega t)}$$

---

* WT: High time!

ET: Stay prepared for the worst. What will come is the best.

We have obtained $\hbar k = p_x$, the x component of the momentum, because of the way we used $\hbar$ as a factor. So the operators for energy and the three components of momentum are

$$-\frac{\hbar}{i}\frac{\partial}{\partial t}\,;\qquad \frac{\hbar}{i}\frac{\partial}{\partial x}\,;\qquad \frac{\hbar}{i}\frac{\partial}{\partial y}\,;\qquad \frac{\hbar}{i}\frac{\partial}{\partial z}$$

The minus sign in $-(\hbar/i)(\partial/\partial t)$ is introduced so that the propagation for positive values of the momentum and energy should occur in the positive x (or y or z) direction.*

Thus we have functions for which at least some properties of particles (energy, momentum) have sharply defined values. In quantum mechanics,† one has for each physical quantity an operator and eigenfunctions for which this physical quantity has definite values.

Why did we use for the eigenfunction associated with momentum an expression containing e^{ikx}, rather than e^{kx}, which would be much easier to visualize? Because (in a coordinate system where x increases as we go to the right) e^{kx} rapidly increases when we go to the right. This must mean that the particle represented by the wave is much more probably to the right than it is here. And no matter where we go, the particle will still be, with practical certainty, even farther to the right. For e^{ikx}, we cannot assert that the particle is at any place, a crucial point which we shall discuss in the next chapter. But e^{ikx}, at least, gives us ignorance about the position of the particle, rather than the absurd statement that the particle must be eternally elsewhere. Even so, the eigenfunction e^{ikx}, by extending to infinity, remains somewhat unsatisfactory.

What is the eigenfunction for the position, described by the coordinate x, where x takes on a specific value x_o? One of the found-

---

* In 1933, I heard Niels Bohr discuss this question for nearly an hour, creating the remarkable illusion that he did not know the answer all the time. He usually discussed questions for which he did not know the answer. If he did know, at least he pretended not to know.

† The obvious name of the new science which turned me away from pure mathematics, which was difficult, and from chemistry, which was easy.

ers of quantum mechanics, Paul Dirac, with his addiction to sharp,
if absurd statements, defined this function as one which depends in
a peculiar way on the coordinate x: it is zero for every value of x
except for $x = x_o$. The function is named for the symbol used to
represent it, δ, and its inventor. So it is called the Dirac δ function.
The operator representing x will be the instruction: "Multiply the
Dirac δ function by x." Then we have $x\delta(x - x_o) = x_o\delta(x - x_o)$,
where δ is always zero except for $x - x_o = 0$. So for $\delta(x - x_o)$
multiplication by x gives the same result as multiplication by x_o.
Thus $\delta(x - x_o)$ is indeed the eigenfunction for the operator x and
the eigenvalue is x_o.

When Schrödinger received Einstein's letter containing the news
about de Broglie, the behavior of wave equations was known to
physicists, the ideas of linear operators much less so. Schrödinger
produced his wave equation by intuition, almost by magic. Looking
at the equation with the idea of operators in mind, we can make
Schrödinger's result appear almost logical.

An electron in a hydrogen atom has the potential energy
$-e^2/r$. The minus sign means that you have to add the energy
e^2/r to remove the electron to infinity. The operator of the potential
energy is: "Multiply by $-e^2/r$." One can argue this in the same way
as we argued the operator for x.

The kinetic energy of the electron is

$$\frac{p^2}{2m} = \frac{1}{2m}(p_x^2 + p_y^2 + p_z^2).$$

The operator for p_x^2 is the operator for p_x applied twice. Indeed,
if we take an eigenfunction of p_x and apply the operator of p_x
$= [(\hbar/i)(\partial/\partial x)]$ then we get back the original eigenfunction multiplied by p_x. Now do it over again and you get the same function
multiplied by p_x a second time. Altogether you multiplied by p_x^2.
That is, p_x and p_x^2 have the same eigenfunctions but the operator
must be used twice; we get $(\hbar/i)(\partial/\partial x)[(\hbar/i)(\partial/\partial x)]$, which is
$-\hbar^2(\partial^2/\partial x^2)$. Here the symbol $\partial^2/\partial x^2$ means that we must differentiate twice in a row by x, keeping the other coordinates constant.

Or in mathematical language, $(\partial/\partial x)(\partial/\partial x) \equiv \partial^2/\partial x^2$. To tell you the last secret of mathematics, $\equiv$ is not an equality (which would be $=$) but an identity or simply a definition.

Now Schrödinger's famous wave equation for the wave function $\psi(x, y, z, t)$ is the statement that the operator for the kinetic energy plus the operator for the potential energy add up to the operator for the total energy. For an eigenfunction of the energy, this must be the energy E times ψ. In one formula

$$-\frac{\hbar^2}{2m}\left(\frac{\partial^2\psi}{\partial x^2} + \frac{\partial^2\psi}{\partial y^2} + \frac{\partial^2\psi}{\partial z^2}\right) - \frac{e^2}{r}\,\psi = E\psi,$$

where E is the eigenvalue of the energy. This is the equation which Schrödinger could write down but could not solve. Not without expert help from mathematicians.

I do not quite believe this story. I believe that Schrödinger solved this equation for the lowest, stable state of the hydrogen atom. Even we can solve it.

Let us first assume that ψ depends only on r, the distance from the attractive center. Then every mathematician and even every physicist knows that $\psi(r)$ satisfies the above equation which can be rewritten* as:

---

* WT: Can every student know it too?

    ET: You are asking for trouble, but here it is. Consider ψ to be like the potential ϕ in Chapter 7. Then $(\partial/\partial x)\phi$, $(\partial/\partial y)\phi$, and $(\partial/\partial z)\phi$ can be considered as components of the electric field. The divergence of this field is $4\pi\rho$ (ρ is the charge density in this region). Mathematically, this means $(\partial^2\phi/\partial x^2) + (\partial^2\phi/\partial y^2) + (\partial^2\phi/\partial z^2) = 4\pi\rho$. Now let us express this same situation for the case that ϕ only depends on r. Then the electric field or the lines of force will have a component only in the r direction and this is $(d/dr)\phi$. We now ask: "What is the difference in the lines of force entering through a spherical surface of radius r and those exiting through a spherical surface of radius $r + \Delta r$?" Those entering will be $4\pi r^2 (d/dr)\phi$. The difference will be obtained by differentiation

$$\Delta r 4\pi \frac{d}{dr}\left(r^2 \frac{d}{dr}\phi\right) = \Delta r 4\pi\left(2r\frac{d}{dr}\phi + r^2\frac{d^2\phi}{dr^2}\right).$$

$$-\frac{\hbar^2}{2m}\left[\frac{d^2\psi(r)}{dr^2}+\frac{2}{r}\frac{d\psi(r)}{dr}\right]-\frac{e^2}{r}\psi = E\psi.$$

I am sure that Schrödinger saw the lowest solution of this equation at once: $\psi = e^{-r/r_o}$. One only must set

$$-\frac{\hbar^2}{2m}\frac{d^2\psi(r)}{dr^2} = -\frac{\hbar^2}{2mr_o^2}\psi = E\psi$$

and

$$\frac{\hbar^2}{mr_o r}\psi = \frac{e^2}{r}\psi.$$

This gives $r_o = \hbar^2/me^2$, the radius Bohr derived in 1914 from his simple and absurd assumptions. Also, the predicted energy remains the same, $E = -\hbar^2/2mr_o^2 = -e^4m/2\hbar^2$. The negative sign means that one has to add $e^4m/2\hbar^2 = 13.5$ electron volts (eV) of energy to separate the electron from the nucleus, the singly charged proton. For nuclei of other atoms carrying an integer multiple of Z of the charge on one proton, one may hope that the same argument will hold as it did for the hydrogen atom, except that the attractive potential $-e^2/r$ must be replaced by $-Z(e^2/r)$. This should give for the binding energy $-Z^2(e^4m/2\hbar^2)$, in good agreement with the binding energy for the most tightly bound electrons in any atom.

By this time Schrödinger must have known that he had an atomic bear by the tail and he went to the mathematicians for the other solutions to his equation. The mathematicians said: "This is easy. If you call the lowest energy R (for Rydberg), then the eigenvalues are $-R/n^2$, where n is any integer. And for each n there are n^2 independent solutions, one of them with a ψ function depending

---

If we divide by the volume of the shell, $4\pi r^2\Delta r$, we get the excess of the lines leaving over those entering per unit volume and that is $(2/r)(d/dr)\phi + (d^2\phi/dr^2) = 4\pi\rho$. The same argument applied to ψ gives the simplified Schrödinger equation.

only on r and the others with eigenfunctions depending on the angles as well."

Even this was not enough. It is quite straightforward to obtain the wave equation describing the behavior of any number of particles, say P, where we enumerate them from 1 to P. Furthermore, the dependence on time was also to be included. The function then depends on $3P + 1$ variables: $\psi(x_1, y_1, z_1, x_2, y_2, z_2, \ldots, x_P, y_P, z_P, t)$. The wave equation is

$$\sum_{j=1}^{P} T_j\psi + V\psi = -\frac{\hbar}{i}\frac{\partial}{\partial t}\psi.$$

We must explain what each term means: $\sum_{j=1}^{P}$ stands for the summation; one has to add the kinetic energy of each particle from 1 to P. The kinetic energy has the same form for every particle. For instance, for the last particle it will be

$$-\frac{1}{2m}\left(\frac{\partial^2\psi}{\partial x_P^2} + \frac{\partial^2\psi}{\partial y_P^2} + \frac{\partial^2\psi}{\partial z_P^2}\right).$$

The potential energy depends on the configuration of the P particles. In atoms and molecules, it is, to a good approximation, the sum of the electrostatic interactions taken over all pairs. Finally, the operator $-(\hbar/i)(\partial/\partial t)$ stands for the total energy of the whole system.

The equation turns out to be quite accurate, missing a few corrections (which are usually unimportant), but also requiring a major and surprising restriction. It is important to explore the reason for this one point not yet occurring in Schrödinger's original proposal.

The general interpretation of ψ and its dependence on the configuration is quite straightforward: the absolute square of ψ, that is $|\psi|^2$, is a positive number giving the probability of the configuration of particles occurring in ψ. In most cases, including the description of a heavy atom, ψ can be written in a reasonable approximation as a product of the individual wave functions of the electrons. Then the probability of the configuration becomes the product of the independent probabilities of finding each electron where its ψ function indicates that it should be. The electrons interact quite strongly, but it turns out to be sufficient to calculate the function of each in the average of the field produced by the other electrons.

Right here we run into big and interesting trouble. The radius of the orbit of lowest energy if the nuclear charge is Ze should be $\hbar^2/mZe^2$, the same as in the hydrogen atom except for the factor of Z in the denominator.* Therefore heavy atoms (those with high Z values) should become rapidly smaller. Actually, they become somewhat bigger.

The solution comes in two steps and, with them, Schrödinger's proposed equation remains valid.

The first corrective step is that no two electrons can occupy the same ψ function. This is accomplished by postulating that the ψ function changes sign whenever two electrons are exchanged. That is $\psi(1, 2) = -\psi(2, 1)$. To make distinction easier, we write ϕ and χ for the wave functions of electron 1 and electron 2. Then the proper approximate solution will be $\phi(1)\chi(2) - \chi(1)\phi(2)$. Now if $\chi = \phi$, then $\phi(1)\chi(2) - \chi(1)\phi(2) = 0$ and we cannot use the function for any two electrons.

A more thorough discussion shows that the two wave functions must be utterly different. For high Z values, the most strongly bound electrons are, indeed, described by a wave function with an approximate radius of $\hbar^2/mZe^2$. But the outermost electron is forced into a high enough energy (that is, high enough n value) to account for the increasing size of heavy atoms in a quite satisfactory fashion.

One interesting and important consequence of the "exclusion" principle (no two electrons in the same state) proposed and explained by Pauli, is that the principle is self-perpetuating—the passing of time does not alter it. If antisymmetry $[\psi(1, 2) = -\psi(2, 1)]$ holds at any time, then ψ is antisymmetric forever. This follows from the fact that the properties of the particles in question, in particular, the

---

* WT: Will that not be changed due to the interaction of the electrons?
  ET: It will, but only by a few percent.

electrons, are absolutely the same. That is, they must all have the same charge, the same mass, and also any property that could exert influence on them must be the same. Then the time-dependent Schrödinger equation, written as $-(\hbar/i)(\partial/\partial t)\psi = \mathcal{O}\psi$ says: If ψ on the right-hand side changes sign upon interchanging two electrons, the expression on the left-hand side will do the same. Indeed, $\mathcal{O}$ stands for the operator which is the sum of the kinetic and potential energies and this does not change when two electrons are swapped. If, therefore, ψ changes sign when two electrons are interchanged, the same holds for its change with time, $(\partial/\partial t)\psi$, and the original property of antisymmetry will not get lost.

This statement that demands absolutely identical particles encouraged physicists that they were close to having found all the tools needed to explain everything—at least in this area they need not search among particles obeying the Pauli exclusion principle for those that might have slight differences. If there *were* differences, then in the course of time more than one particle would have slipped into the same orbit as another, almost identical twin.

But I have said that two steps were needed to save Schrödinger's equation. The second comes when one finds not one but two electrons in the lowest possible energy state of $-E = Z^2(e^4 m/2\hbar^2)$. And the first state above that, which according to Schrödinger's mathematical friends has the energy $-R/4$ and has four ψ states, contains not four but eight electrons. Indeed, for $Z = 2$, we find two electrons present and both are happy to be in the lowest energy orbits, filling what we call the K shell and giving us the inert gas helium. It is an inert gas because the electrons are quite satisfied with their low energy and don't want to be rearranged in any interaction with electrons from other atoms. Similarly, when we go to an atom with eight more protons in the nucleus and eight more electrons surrounding it, filling the states with energy corresponding to $-R/4$, the L shell is filled and this atom with $Z = 10$ is again an inert gas, neon. Of course, the energy is not a factor of $1/4$ lower than in helium, because Z has increased and actually the neon atom is a bit bigger than the helium atom. With these two examples we have the beginning of the Periodic Table of the elements, first pointed out by Dmitri Mendeleev in 1869, in which the chemical properties of the elements repeat. For instance, the inertia of helium is repeated in neon.

Why can the two electrons coexist in the same state? The answer is that electrons have one additional, almost hidden property. They have "spin." In any state an electron can have imposed on it an angular momentum to the right, or an angular momentum to the left. And this tiny angular momentum has the size of $\hbar/2$.* In most cases, its only effect comes from the fact that electrons of opposing spins of $+\hbar/2$ and $-\hbar/2$ like to get paired, so that apart from spin, there can be two electrons in the "same" state.

In a greatly oversimplified discussion, this can be called the reason for the chemical bond, the situation where two electrons are found in a wave function that belongs not just to one atom but to two neighboring atoms. In a more detailed investigation, the spins can be found and measured. In some exceptional materials, like iron, quite a few electrons will have their spins lined up. The spinning motion of the electron charge produces a magnetic field, and this lining up of spins is the explanation of the spontaneous magnetism or ferromagnetism of iron.

Now has my simple theory of quantum mechanics become complicated enough?† It has to be this complicated because it is supposed to allow us to calculate accurately all of the facts of chemistry and physical chemistry, and thus produce all the facts about the structure of matter that we encounter in everyday life. The physics involved is simple, a statement I make because Schrödinger found

---

* WT: But earlier we saw that $E = \hbar\omega$, so that $\hbar$ has the property of energy divided by frequency or energy times time.

ET: Yes, but energy has the dimensions of mass times velocity times velocity, as in kinetic energy, so $E \sim MVV$. And energy times time must have dimensions of $MVVT$, but VT is a length, L, so $\hbar$ has dimensions of MVL, or the same dimensions as momentum times length. But angular momentum is momentum times a lever arm. In Bohr's theory the angular momentum of the electrons in their orbits was quantized; it had to be a multiple of $\hbar$. For electrons, each possible increment of angular momentum is indeed $\hbar$, but the lowest possible angular momentum is not zero but $\hbar/2$.

† WT: With the differential equation of Schrödinger in almost countless dimensions and with the almost unobservable spin, I can, for the first time, say honestly, yes!

on his first try the precise equation that *does* describe the behavior of matter precisely—in those few cases where a precise mathematical solution is possible. And there is the difficulty. Most of the interesting problems of chemistry are far beyond the point where mathematical solutions can be obtained. The solutions are beyond the capability of our computers. That we can make only slow progress on this incredibly involved *terra incognita*, I blame on the complications of mathematics. What can be said in reasonably simple terms, I shall attempt to explain in Chapter 12.

But first I want to discuss a surprising and important consequence of dualism: the future is uncertain.

QUESTIONS

1. In the ionosphere (the upper atmosphere containing free electrons) the phase velocity of radar waves is given by the formula $c(1 - \alpha\omega^{-2})^{-1/2}$, with α a constant. This is greater than c. What is the group velocity?

2. Formulate in terms of the Schrödinger wave equation the idea that, in a molecule, the electrons move in the field of fixed nuclei while the nuclei move in the average field of the electrons. To what extent does this formulation fall short of an exact description of molecular behavior?

Chapter 11

THE UNCERTAINTY PRINCIPLE

In which it is shown that you cannot know everything (position and momentum) about anything, and in which the authors attempt to extricate themselves from the philosophical consequences.

When we discussed the wave–particle dualism, we saw that particles and waves behave in the same manner to a surprising extent. But, of course, they still are completely different.

Actually we associate with particles a set of ideas taken from everyday experience. We think of a particle as indivisible, at least as far as the immediate application of the concept is concerned. A particle is not partly here and partly there. Whenever we get to a quantity that is not to be divided any further in the problem we have at hand, we speak of a particle. A molecule is a particle as long as it is not dissociated; an atom as long as it is not ionized; a nucleus as long as it is not fragmented.

Of waves, we have the idea that—like waves on water—they extend in a continuous fashion. The water is, of course, not essential.

Neither is (and this may seem strange) the presence of any other substance. The classical example was light. "Ether" was postulated as the substance in which electromagnetic waves propagate. But neither the "etherwind" nor any other manifestation of ether could be observed. In the end, the ether (like the Cheshire Cat) vanished and only the waves remained. How, indeed, could matter be explained by waves if these in turn required material in which to propagate?

We further assume (although this is only approximately true in macroscopic physics) that if two wave processes can occur in nature, their sum may also occur. Light waves may be added to each other. So can Schrödinger waves. This is called superposition or interference and this phenomenon is characteristic of waves. Where wave maxima coincide and reinforce each other, we speak of positive interference. Where wave maxima and minima coincide and tend to cancel, we talk of destructive interference. Therefore (at least at one special location) something plus something may give nothing.

Furthermore, we do not merely talk of sine or cosine waves. Any function in space will do, like the functions shown in Figures 1 and 2. Each can be written as a superposition of sine or cosine waves. The mathematicians have practiced this for a long time and called it Fourier analysis of a function. (The mathematicians seemed to have thought of everything first.*)

If we have a wave process as shown in Figure 1 and call the function $\phi(x)$, then the probability of finding the corresponding particle at the position x will be $|\phi(x)|^2$. Furthermore, $\phi(x)$ can be obtained by superposing waves of wave numbers $k_1, k_2, \ldots, k_n$, $\phi(x) = a_1\psi_1(x) + a_2\psi_2(x) + a_3\psi_3(x) + \cdots + a_n\psi_n(x)$ (in reality we need a continuous array of k values), with amplitudes $a_1, a_2, \ldots, a_n$. Then the probability of finding the corresponding momentum

---

* ET: There are exceptions: Newton, a physicist, invented differentiation all by himself.
  WT: He did it while he was young, intelligent, and a mathematician. Later he decayed into a physicist, then an alchemist, and finally into an administrator.
  ET: But he saved his soul by publishing his *Philosophiae Naturalis Principia Mathematica*.
  WT: And you see what the last word is?

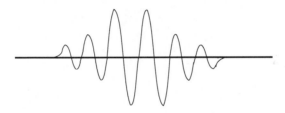

Figure 1. This wave function can be formed by adding together sine or cosine waves.

values $\hbar k_1$, $\hbar k_2$, ..., $\hbar k_n$, will be $|a_1|^2$, $|a_2|^2$, ..., $|a_n|^2$. For this to make sense, it must be true that the sum $|a_1|^2 + |a_2|^2 + \cdots + |a_n|^2 = 1$, because the probability for the particle to have some momentum must be 1 = certainty. Fortunately this is true if $\int |\psi(x)|^2 \, dx = 1$, that is, if the probability of the particle to have some position is 1 = certainty.*

Throughout the discussion of physics we have used the concept of probability. It is necessary to distinguish two quite different uses of this concept. In statistical mechanics we use statistics because we are lazy. In quantum mechanics we use it because we cannot do otherwise. In the first case, probability is a convenience. In the second case, it is (according to the Niels Bohr school) the only way to escape a clear-cut contradiction.

At the time of the great depression in the early 1930s Eddington expressed this difference in an unforgettable manner. He likened the procedures in statistical mechanics to the use of paper money. It is more easy to handle than gold. But if anyone doubted the value of the paper currency, he could any time go to the bank and get gold for it. The paper of statistical mechanics was solidly backed up by

---

* WT: Where does this miracle come from?
ET: Of course, from the mathematicians. Even worse, in this case two French mathematicians are involved. One is Fourier, who helps us in the case of eigenfunctions of k or p, the other is Hermite, who has selected among the linear operators those where the probabilities always add up to one for any set of eigenfunctions, as in the case we discussed.

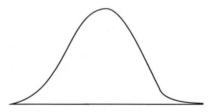

Figure 2. It is also true that a wave function such as this can be formed from sine and cosine waves.

the golden value of physics firmly planted in causality. What happened in quantum mechanics was physics went off the gold standard. We can rely only on probabilities. There is no deterministic system to back up our theory.

One must say all of this in a more clear-cut manner. In quantum mechanics we have two stages. We can make calculations in which wave functions change according to the laws of differential equations and act in accordance with the straightforward concept of causality. But there is another stage, the performance of measurements. If an electron is represented by the wave function shown in Figure 1, and if we want to determine its position more accurately than the spread shown in the figure, we can do so. But there is no way whatsoever to predict more than a probability of the outcome of such a measurement.

It is important to make clear what is simple and what is difficult about this application of the probability concept.

It is simple because we have now a way to reconcile the wave concept and the particle concept. If we have a pure sine wave extending to infinity, then the particle represented by the wave has a sharp momentum value given by the wave number. On the other hand, the position of the particle is completely uncertain. Consider now a localized function consisting of a narrow bump, where the position of the particle is well-defined but the momentum (according to the decomposition into sine waves in the Fourier analysis) is quite uncertain. We can change from the one situation to the other by making a measurement subject only to the laws of probability.

Let us look at the same situation on an astronomical scale. A star located a hundred light years away emits a particular light quantum. It spreads out on a hemisphere (the light quantum could have gone only outward from the star, which is one-half of all possible directions) with a radius of one hundred light years. Then an astronomer catches the light quantum on his photographic plate (which of course has an exceedingly small probability) where it starts the growth of a little silver crystal which, when the plate is developed, appears like a small black dot. Just before the light quantum has been trapped, its wave function (which represents an electromagnetic field) has been spread out over a region greater than the universe known to Tycho Brahe. According to Einstein, no process governed by causality can spread faster than the speed of light. Could any process describable by a differential equation (which, of course, must not contradict relativity) thus contract the wave function from cosmic dimensions to a grain on the photographic plate!?! Yet the process occurs in a time short compared to the twinkling of an eye.*

Therefore the measurement process cannot be described by a differential equation or any other way which is subject to the laws of causality. Measurement is tied to probability, which is basically antithetical to causality. The gold of causality is still there whenever we use differential equations. But the gold standard, the principle that in the end everything must be referred to cause–effect relations, is gone forever.

What is difficult about this concept is the understanding of a "measurement." One must realize that such a measurement cannot be described by any differential equations. The measurement creates a new situation. That this can happen outside the strict ideas of causality brought about a famous remark by Einstein: "I can imagine that God governs the world according to any set of rules, but I cannot imagine that He is playing at dice."

---

* WT: Any decent star emits many quanta, even in a very short time.
  ET: Freedom of thought is important and a "thought experiment" is important to eliminate contradictions in one's thoughts.

There are two rejoinders: "Why not?" and "How else can the particle–wave contradiction be reconciled?"

First we address "Why not?" The philosophical reasoning behind Einstein's argument for causality is this: Without causality the link between the outside world and my perception of the outside world would be broken. No causality; no knowledge.

Einstein's argument is simple and plausible—and weak. Even statistical mechanics (paper money on the gold standard) disproves it in a practical sense. The behavior of gases and, more generally, all phenomena governed by heat produced solid knowledge even though it would have been completely impractical to go back to the causal relations which we have expected to lie behind the statistical theory of heat.

Einstein knew that; this is the reason why he introduced God in his statement. He firmly believed in the proposition that even if everything is not known (far, far from it), everything must be knowable. But is it? Quantum mechanics sets definite limits to omniscience. It raises the question whether omniscience is conceivable or even logically consistent. This is much too difficult a question for us to settle here. But the limitations of actual knowledge and the possibility that the knowledge may be limited suffice in principle. I believe that science is possible without the reassuring gold standard of causality.

An oversimplified but still valid summary of the last paragraph is the obvious statement: An answer does not exist to every question that can be asked (possibly not even if the question is free of self-contradictions).

To the second question, "How else can the particle—wave contradiction be reconciled?" I have no answer. I cannot prove that there is no other way. Bohr refused even to consider the question. He was too full of his own answer. It is worthwhile to consider that answer in somewhat greater detail.

An important part of the probability argument was formulated by Heisenberg and is known as the Heisenberg uncertainty principle. The ideas and formulae which recur in Heisenberg's uncertainty principle are similar to those discussed above. Let us first try to use the superposition of several waves with a range of wave n imbers between k and $k + \Delta k$ to localize the wave function as well as possible and to construct a wave packet, such as shown in Figure 1, which extends only to a minimal distance. The corresponding sine waves would be in the range sin (kx) and sin $(k + \Delta k)x$ or sin $(kx + \Delta kx)$. Near $x = 0$, these sine waves (or the similar cosine waves) reinforce each other, since they are almost in the same phase. What happens if you move to a distance Δx? Then the phases are $k \Delta x$ and $k \Delta x + \Delta k \Delta x$ with the phase difference $\Delta k \Delta x$. If that quantity becomes equal to π, then the two waves are in opposite phases and will cancel. Heisenberg concluded that with a range of wave numbers equal to Δk one cannot localize the wave function better than $\Delta x = 1/\Delta k$ or that $\Delta k \Delta x \geq 1$.*

If we now multiply by $\hbar$ and remember that $\hbar k = p$, we obtain $\Delta p \Delta x \geq \hbar$. This is the Heisenberg uncertainty relation. The more you know about p the less you can know about x and vice versa.

The important point about the uncertainty relation is its connection with the process of measurement. What is there in that process which makes it impossible to measure both p and x accurately? After all, it is assumed that either of the two can be determined and can be measured with any desired accuracy.

In fact, there is no limitation in the measuring process itself. Rather one has to assume to begin with that the Heisenberg uncertainty principle is universally valid. If there were any object for which the simultaneous determination of p and x would be possible to higher accuracy, one could use that object to make measurement on any other object to a similar accuracy. But if the measuring rod in itself is inaccurate, the same will hold for any results of measurement. The uncertainty principle is valid universally or it is not valid at all. The discussion of these experiments does not prove the uncertainty principle. They only show that it is a self-consistent postulate.

---

* WT: Why not $\Delta k \Delta x \geq \pi$?

ET: Up to $\Delta k \Delta x = 1$ there is reasonable reinforcement of the waves. When we talk about k we always get into some difficulty about a factor such as π.

This point has been discussed in many specific examples. We shall not occupy ourselves with details. It is plausible that accurate measurement of x of an electron will destroy knowledge of the momentum of the electron, even if the momentum had been known before the position is measured. In fact, one may use for the measurement a light quantum for which the uncertainty relation holds. When the position is measured an uncertain momentum is imparted to the electron.

It is a little more difficult to see how knowledge of the position of an electron is lost when its momentum is measured by scattering a light quantum off the electron. The measurement may be made by determining the frequency shift in the scattered light due to the Doppler effect. In the scattering process momentum is exchanged between the electron and the light quantum. This in itself need not worry us, because it can be taken into account. But at the moment of scattering, the electron changes its velocity and we cannot know precisely when the scattering occurred, because we need time to establish the frequency shift of the scattered light to the required accuracy. A velocity change at an undeterminable time results in an uncertainty in the resulting position, sufficiently large to satisfy the uncertainty relation.

We have not proved the uncertainty principle by discussing experiments; we have only shown its consistency. Could we get along without the uncertainty principle?

No! We need the wave–particle dualism to reconcile the observation of light quanta with Maxwell's theory and to formulate a wavelike theory of the particles (including electrons) of which matter is made. We shall discuss the question of how the uncertainty principle prevents us from cross-examining a light quantum (or an electron) and make it confess whether it is a wave or a particle. If we could get a confession, this would amount to a direct contradiction to wave theory (or particle theory). Yet, we need both.

In Figure 3 we show a plane wave of wavelength λ shining in the x direction on a single slit in a screen and a light sensitive screen behind it. The light sensitive screen might be a photographic plate. If the size of the slit is not very wide compared to the wavelength of

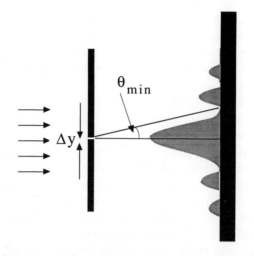

Figure 3.   A plane wave shining onto a screen through a narrow slit will produce a pattern of intensity similar to that shown here.

the light, $\Delta y \leq \lambda$, then there will be a spread of exposure of the plate. A diffraction pattern will arise because of the spread in the y component of momentum of the wave as it passes through the finite slit.

Notice that because the screen blocks the light, and as a part of the beam passes through the slit, we know where the wave passing through the slit is in the y direction. The first minimum of the diffraction pattern is, roughly, $\theta_{min} \approx \lambda/\Delta y$. But for the light to be able to reach that point on the light sensitive screen, $\Delta k_y/k \approx \theta_{min}$, or $\Delta k_y/k = \lambda/\Delta y$. This means that $\Delta k_y \, \Delta y \approx \lambda k \approx 1$, but multiplying both sides by $\hbar$, we get $\Delta p \, \Delta y \approx \hbar$ for the light quantum at the instant it passes through the screen. The broad diffraction pattern is not in contradiction with the particle picture if we take into account the uncertainty principle for the light quantum.

The diffraction pattern exists; it cannot be eliminated. Making the slit narrower merely increases the vertical spread in momentum; making it wider reduces the uncertainty in momentum, but we lose knowledge of position.

Many attempts have been made to cross-examine light quanta or electrons, but the uncertainty principle always provides a loophole just large enough to avoid straightforward contradiction.

Bohr and Heisenberg carried such cross-examination to a merciless third degree.

"Where did the light get the extra Δk_y and the extra momentum $\hbar \Delta k_y$?" they asked. "It must have obtained it from the screen. So let us measure the momentum of the screen before and after the light quantum passed the slit but before the light quantum arrives at the fluorescent screen. Then it will be possible to calculate where, within the diffraction pattern, the light quantum arrives. The probabilistic diffraction pattern would be destroyed and the unique particle nature of the light quantum would be established."

For duality, this would be intolerable.

But Bohr and Heisenberg did not let the light quantum in the diffraction system get away with such an answer. You can't measure the momentum of the screen without losing knowledge of the position of the slit, which follows from the application of the uncertainty principle to the screen or, rather, to the slit in the screen. That measurement must be carried out before the light quantum hits the luminescent screen, if any prediction is to be made. Loss of knowledge of the position of the slit just suffices to prevent the prediction as to where the light quantum will land.

I have given this argument (without all the details) as an example of the discussion that went on in the circles of Bohr and Heisenberg for a couple of years. And there was good reason. Duality and the strict predictability (or rather the limits between predictability and unpredictability) of the future was at stake. The unification of physics and chemistry by quantum mechanics was a great accomplishment. The recognition that our present state of the world exists on the limits of a known past and an unknowable future is a greater accomplishment.* All of this is due to the fact that the uncertainty

---

* WT: I thought that I would know the future when I grew up.

ET: That, too, is what I thought when I was a teenager and, indeed, people used to believe that they could find out the future, at least in principle. The great news is that

principle gives us the latitude to avoid direct contradiction between statistical behavior and the application of causality when the causes are known.

The consequences of the Heisenberg uncertainty principle in terms of our outlook on the world are immense. Before Heisenberg, we saw the past as fixed and immutable and the future determined, mechanistic. After Heisenberg stated and fully discussed his principle, we now view the past as fixed, still, but the future is uncertain. It is in the process of creation. According to 19th century physics, the world was created and then ran in a pattern which was forever immutable. When a hundred years ago, the famous Arab tent-maker–astronomer–poet's (Omar Khayam's) verses were recreated, they read:

> With Earth's first clay, they did the last man knead
> And there of the last harvest sowed the seed
> And the first moment of creation wrote
> What the last day of reckoning will read.

God had finished His creation and He rested—not for the seventh day, but for evermore. He was, simply stated, unemployed.

Today we know that the world is different. It is newly created every microsecond by every atom, every star, and every living being. And what is the role of God? Perhaps He is the conductor of the strange universal harmony and He is the repository of everything we do not know. This is, of course, an irresponsible guess. But the role of the scientist is clear. He is attempting to read the score of the orchestra called the universe.

No attempt to disprove the wave–particle interpretation of quantum mechanics succeeded in the discussion of atomic phenomena. A different kind of objection was raised by Schrödinger. He proposed to amplify the quantum phenomena so that their effect

---

we have found out that knowing the future leads to inconsistencies. In the three statements "entropy can never decrease," "the speed of light can never be exceeded," and "the future can never be accurately predicted," the word "never" might be replaced in the first by "hardly ever" (unless you believe in all of the definitions of Gibbs), but in the second and third statements the word "never" is better replaced by "Never."

should be felt in macroscopic physics and then he pointed out the absurd consequences.

He proposed quite a concrete example; an arrangement is shown in Figure 4. The arrangement consists of a box within a box and a cat in the inner box. The cat is provided with food and milk and oxygen. In the outer box, however, there is a deadly hydrogen cyanide atmosphere. In this outer box there is also a radioactive source and a counter, both well isolated from each other by a shutter. Once— and only once—an alarm clock removes the shutter for one second. The apparatus is so adjusted that there is a 50% probability for the counter to be triggered during that period. Otherwise the counter will be inactive. If the counter is triggered, an electric contact is established which connects a battery and burns out a portion of the screen protecting the inner box. If that happens, the hydrogen cyanide gets to the cat and kills it. Then one waits for a week. At the end of that period, the "wave function" of the system will be a superposition of two states. In the one, the cat is dead, in the other, alive. At the end of the week the apparatus is opened and the act of observation

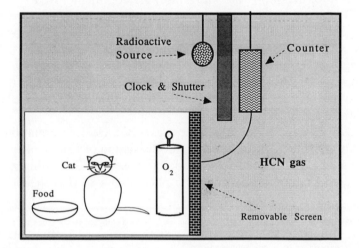

Figure 4.   This is a sketch of the famous "Schrödinger's cat" and his quarters.

brings about one of two situations. The wave function "dead cat" may be established or the cat may be half-resurrected from its half-dead state and is found alive.

In principle, Schrödinger has correctly reproduced the arguments of the uncertainty principle. Prior to the observation neither the dead cat nor the live cat represented full reality. It is the interaction of the observation with the cat system which brings about the reality of either alternative.

Schrödinger then remarks that all of this is certainly nonsense. But the argument of Schrödinger is oversimplified. When we talk about a wave function in the final situation, we are really considering a wave function in 10^{30}-dimensional space (more or less). Interference between the two states "dead cat" and "live cat" can occur only if one considers configurations in the two wave functions which agree in all of the 10^{30} coordinates. If even a single electron is in a different spot, interference does not occur. The chance of such an interference is incredibly small. The famous example that ten monkeys using ten typewriters should produce by chance all the works of Shakespeare is an excellent bet by comparison.

But if there is no interference, then the quantum description is essentially the same in its results as the classical one. Whether in the fateful second a particle has been sent into the counter can be established and that is a reasonable occasion on which to make the observation. From there on chains of causality will operate and it makes no difference in the results whether we wait for a microsecond or a century before we make the measurement and decide the alternative.

The result of a discussion of the act of measurement remains that we must distinguish the causal developments described by the differential equations and the measurement, which gives a statistical—and therefore an unpredictable—result. At what stage we make the measurement does not matter from the time onward when the alternative possibilities are separated into nonoverlapping "wave packets." And it should be emphasized again that two wave packets are nonoverlapping as soon as there is at least one electron, nucleus, or light quantum on which the two wave packets differ. An overlap is not established unless you consider the same configuration down to the last particle.

Classical physics remains an indispensable part in the understanding of quantum mechanics. It is classical physics which provides the direct objects of our experience. Our concepts and our words relate to classical physics. We may talk in a consistent manner about quantum mechanics by itself. But we do not know what we are talking about without reference to classical physics.

Einstein said that if there were no causality, there could be no science. Bohr said that if there are no words, then there can be no understanding. Bohr's statement has something of the flavor of Einstein's, but it is more modest. He was pointing out that without words, we cannot talk—or think. Therefore, we have to start with the thinking process as we have it. Only then can we discuss its limitations.

I conclude that we cannot start our studies with quantum mechanics. A bridge between quantum physics and classical physics is necessary and that bridge is the classical measurement. You may place that bridge anywhere between object and observer, but it is most practical to place it as close to the object as possible. That means that the measurement may be done and should be done as soon as the probability of interference phenomena has become zero.

When can we disregard interference? In a way the answer should be "never." If an originally simple atomic process, in which quantum theory is needed and interference occurs, results at a late stage in a complicated configuration, one always can reconstitute the original situation if we could invent a way in which the velocity of every single particle is reversed. This, of course, is completely impractical and runs counter to the second law of thermodynamics, the law of increasing disorder, introduced in Chapter 6.

So we have to ask again: When is a reversal to the original state impractical? One good answer is that an "irreversible" process in classical physics makes a return to the original state thoroughly impractical. The very word "irreversible" expresses the idea. The reason for the irreversibility is that disorder is created and the probability that order should be automatically reestablished is exceedingly small.

In the case of Schrödinger's cat, the functioning of the counter is irreversible. In a photographic plate the formation of a little silver crystal is irreversible. Something of the same kind occurs in every practical measuring process.

It is wrong to believe that the difference between quantum mechanics and classical theory is in the size of the objects which we consider. The real difference lies in the question of order. When the order is perfect (which hardly ever occurs on a big scale) and when uncertainties have been reduced to the limit given by Heisenberg, quantum theory applies. But if there is enough disorder, we soon are allowed to discuss the world in terms of common experience.

But at the same time we should never forget that in the narrow confines of quantum mechanics the world is reborn in every instant. The two revolutions of Copernicus and that of Bohr are in a way complementary. Copernicus demonstrated (in the end) how infinitesimally small we are. Bohr reminds us that, though the universe is overwhelming, we are, in our continuing action, only partially dependent on the past. We retain an irreducible element of freedom. And, according to Einstein, these actions (whose freedom he has wrongly denied) spread their effects with a finite velocity, the tremendous speed of light, into the universe. It is not only the universe that is almost infinite; the consequences of the actions of every human—or every fly—also border on infinity.

QUESTIONS

1. Let us consider two screens A and B with a hole in each screen. Let us measure the time at which an electron passes through the hole in A and the later time at which it passes through the hole in B. These measurements can be carried out with arbitrary accuracy (though at the time of the passage through A, one cannot predict the

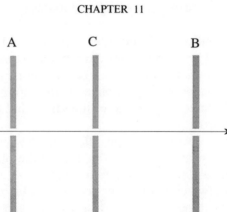

time of the passage through B). Knowing locations and times at two points (and assuming no forces between the two screens) one can say that between the two points the electron must have moved on a straight line with a constant velocity. Therefore, one can state with as great an accuracy as one pleases what the position at an arbitrary intermediate time was (at C in the figure) and what the velocity— and therefore the momentum—values have been at the same time. Is this not a violation of the uncertainty principle?

2. Einstein suggested a measurement that would determine with unlimited accuracy both E and t, thus violating the uncertainty principle in the form $\Delta t\, \Delta E \geq \hbar$. He took a box with perfectly reflecting walls with a light quantum trapped inside. A clock can open the shutter in the top for some precisely determined interval Δt, allowing the photon to escape. By weighing the box before and after and using $\Delta E = c^2 \Delta m$, and then looking at the clock to see what Δt had been,

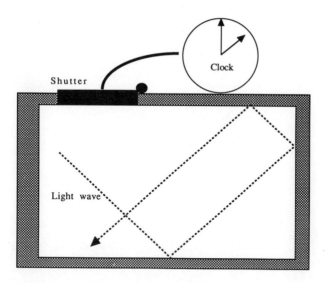

he would have ΔE and Δt independently and both could be made as small as desired.

It took Bohr a (sleepless?) night to find the flaw. What was it?

Chapter 12

USES OF NEW KNOWLEDGE

*Where the reader learns by some examples how
the incredible can turn into the commonplace*

We and our computers are too clumsy to solve differential equations
in one-hundred-dimensional space, or to find the correct combination among ten thousand eigenfunctions that represent the detailed
state of matter that is jointly studied by chemistry and physics. This
is why we shall need scientific as well as industrial laboratories for
the next thousand years.* Since we can't figure out how to play a
perfect game of chess, it is not surprising that a bunch of gallium
and arsenic nuclei, together with their electrons, can play some unexpected tricks on us.[†]

---

* WT: Are you sure of the time scale?
   ET: No, but it is likely that the right number is closer to fifty years than to a million.
[†] WT: What about some carbon, hydrogen, oxygen, and nitrogen nuclei?
   ET: Aha! You are interested in yourself. So am I.

Yet, although we can't calculate exactly, at least we can approximate. And approximation can lead to a qualitative understanding, which in turn guides experiments that lead to applications. The number of possible examples of this transition from basic discoveries to technologies to useful products is almost infinite. We shall discuss at first the behavior of electrons in a solid. Using more explicit words, we shall talk about conductors, semiconductors, superconductors. Finally, we shall talk about lasers. Unfortunately, we can say too little about the most interesting substance: protoplasma.

In a hydrogen atom, an electron can have negative energies, $-R/n^2$, where R is the Rydberg constant and n is an integer. For positive energies, the electron gets loose and at great distances from the nucleus, its eigenfunctions become $e^{i(k_x x + k_y y + k_z z)}$ with an energy $(\hbar^2/2m)(k_x^2 + k_y^2 + k_z^2)$, so that any positive energy is permissible. At closer distance, the wave functions become more complicated; they describe the scattering of electrons by protons. The precise wave functions are then not as simple, but are well known.

For heavier and solitary atoms containing more than one electron, the precise wave functions and energies cannot be obtained by calculations. But it is still true that, for bound electrons, only certain, sharply defined energy eigenvalues are permitted; free electrons can possess any energy. In the solid state, electrons behave quite differently.

I restrict our attention to the simplest and quite widespread form of solids, that is, to crystals. In a crystal the arrangement of atoms in a limited region is repeated again and again, in all three dimensions. That limited region is called a "cell." We refer to the repetition as "periodicity." If one lets a single electron move in a crystal, the general behavior is different from that found in atoms. One should now consider the behavior of the wave function of the electron in one cell. A simple behavior for the electron in the crystal is that it should repeat that behavior in every cell and join those wave functions together at the boundaries of the cells. In that way, we get one state, the equivalent of the orbit in the solitary atom. According to Pauli's exclusion principle, that state can hold one electron—or rather two, if one takes spin into account. Where do we put all the other electrons that should be present, even if we restrict our attention to a small volume of the crystal?

The crystal is built up, as we have stated, from a repetition of cells in three dimensions. Depending on the kind of crystal, these three directions may or may not be perpendicular (the mathematicians say "orthogonal") to each other. Whether or not they are perpendicular, we shall call the three directions, x, y, and z. We count the cell by the three integers, n_x, n_y, and n_z, which count the number of moves we made in the x, y, and z directions. Then we add the wave functions in the cells with phase differences so that we obtain for the phase at n_x, n_y, and n_z, the factor $e^{i(k_x n_x + k_y n_y + k_z n_z)}$. Here k_x, k_y, and k_z are the same as the wave numbers we used to have, except that they are multiplied by integer n values so that $0 \le k_x < 2\pi$. (Note that we don't include $k_x = 2\pi$, because n_x is limited to integer values and $e^{2\pi i} = 1$.) What we now have is rather different from the old momentum wave function, because the exponential factor only compares the wave functions (or their phase) of the cells. Inside the individual cells, the wave function will be more complicated. Indeed, their behavior will vary with the values of k_x, k_y, and k_z continuously and in such a way that for $k_x = 0$ and $k_x = 2\pi$ the behavior is the same (and similarly for k_y and k_z, of course). In the region formed by $0 \le k_x < 2\pi$, $0 \le k_y < 2\pi$, and $0 \le k_z < 2\pi$, the energy varies continuously; this region is called a Brillouin zone.* Various Brillouin zones have various energies which may or may not overlap. So here is the simple result of this discussion: for electrons in crystals within the Brillouin zones, a continuum of energies is permitted. But there may be energies not in these zones and no electron can have such energies.

---

* Louis M. Brillouin made this important contribution of identifying this behavior before World War II. I met him during the war when he had fled France to the United States. He worked for IBM (which used to have signs displayed in all of its company offices that read: THINK!). I asked him how he liked working for IBM, and he expressed great satisfaction. I thereupon asked him whether he had any sign on his desk. His eyes lit up and he said, "Oh, yes. It says 'réfléchissez!' " I agreed with him. It is not sufficient to think about the Brillouin zones; you have to reflect on them.

The Brillouin zones behave quite differently in different energy regions. For a lowest-energy electron in a heavy element, such as gold (that would be, therefore, an electron in the K shell of gold), the wave functions near one gold nucleus and near the neighboring gold nucleus hardly overlap. The consequence is that the Brillouin zone has an exceedingly small energy width. The energy is almost the same as the sharply defined energy in an isolated gold atom. For higher energy levels, the Brillouin zones become broader. But the width of the regions of forbidden energy still remain greater than the regions of permitted energies. In the end, we come into the interesting region of the highest energies, the outermost electrons. There the width of the Brillouin zone and the width of the forbidden region become comparable. Here, for the outermost electrons, we can have two extreme cases.

The first extreme is the case of the noble gases (in their solid states*), where there are sufficient electrons to fill a Brillouin zone and where it will take a considerable amount of energy to move an electron into a higher state. In this case, electrons, in general, won't flow and you have an insulator. The other extreme case is that of a metal, such as sodium or gold, where the outermost electrons suffice to fill half (or some other fraction) of a Brillouin zone. Then an infinitesimal amount of energy suffices to move electrons around, to bring about a current. We then have a conductor.

It is worthwhile to consider both of these cases a little more closely.

One important kind of insulator is a salt crystal, like sodium chloride. Even in the solid state, sodium and chlorine are better considered to be present as ions, rather than as atoms. Indeed, positive and negative ions are closely packed and have nowhere to go as we have seen in Figure 1, Chapter 8. The Brillouin zone of highest energy is filled by electrons around the negative chlorine ions in a configuration similar to the electron configuration in the rare gas, argon. Sodium chloride is colorless. Indeed, the single crystal in the pure state is transparent, because to get to the next highest Brillouin zone

---

* Yes, a gas becomes a solid at low temperatures.

requires a higher energy and frequency ($\hbar\omega$) than is available in the visible light. The crystal is a good insulator. If one adds an electron to the crystal, it can and does carry electricity for some distance, but not far. The positive sodium ions will be attracted to the electron; the negative chlorine ions will be repelled. One may say that the electron had dug a trap for itself, a region around which the ions are appropriately displaced so that the electron will not want to move out of this trap again.

An insulator of an entirely different kind would be a diamond crystal in which each carbon is surrounded by four neighboring carbons. A pair of electrons, described as a chemical bond, is shared by any two neighboring carbons. All of these electrons, eight of them near any carbon atom, fill a Brillouin zone. Diamond is an excellent insulator and is again transparent. One needs even more energy than in the case of sodium chloride to lift an electron into an empty Brillouin zone.

By contrast, lithium, to take a truly simple case, has one more electron than helium. (In helium, for the first time, electron pairs fill a very stable orbit, that of the K shell, near the nucleus.) Lithium, because it has that one more electron and considering the two possibilities because of the spin, will have a Brillouin zone that is only half-filled. These electrons in this highest partially filled Brillouin zone are shared by the lithium atoms and the lithium lattice is held together by them. At the same time, they can accept minimal energies. The electrons are mobile and we have a conducting metal.

In principle, the difference between metals and insulators is clear. In the first, electrons can accept small changes in energy. Indeed, they absorb electromagnetic waves of long wavelength, the shorter infrared lengths into the red, and the even shorter into the visible. They do absorb almost indiscriminately, but in addition to their capability of taking up energy, we should ask how quickly they lose that energy. They do lose it by interaction with lattice vibrations. This is a slow process and, therefore, metals should be excellent conductors of electricity.

This is, in fact, just what they are. By wrapping insulating material a fraction of an inch thick around a reasonably thin 100 mile wire, the electrons will run the 100 miles along the wire rather than cross the insulator. Further, this remarkable high conductivity continues to increase as the temperature is lowered and vibrations decrease. In theory the resistance finally drops to zero at absolute-zero temperature. There is a condition that has to be satisfied, however: the metal must be pure. The free motion of the electrons is due to the strict periodicity of the lattice. The current can be impeded by an impurity which disturbs the periodicity. Thus, at absolute zero, irregularities of the lattice cause a resistance similar to the distortion of the lattice due to lattice vibrations, which become strong at high temperatures.

Even without detailed calculations, the explanation of the structure of matter by quantum mechanics describes the properties of metals and insulators in simple and straightforward terms. But how does one split the difference between metals and insulators?

Assume that electrons manage to fill Brillouin zones up to the last one. But the next Brillouin zone, which is empty of electrons, can accept electrons provided only a small, but definite, amount of energy is added. Then we have what is described as, in entirely appropriate terms, a semiconductor. (You also can get a semiconductor if, in the neighborhood where the electrons have the highest energy, two Brillouin zones slightly overlap.) These semiconductors play an extraordinary role in electronics.

From the time that Thompson discovered the electrons, these particles invited experimentation and application. At first, the game with electrons was played under the most simple of circumstances in vacuum. What is new and interesting about the preceding discussion is that a regular periodic lattice behaves almost as a vacuum. Almost, but not quite. On the interface between two semiconductors, or between a semiconductor and a metal, one can find or introduce conditions so that electrons can pass one way, but not the other. In other words, with semiconductors we can construct valves for electrons. The famous transistor, which makes modern electronics possible, is just such an electron valve.

With the help of transistors and other tricks, electronics can operate a million times—perhaps even a billion times—faster than clumsy macroscopic machinery. That is why the modern computer machine has deprived an old saying—"as fast as thought"—of all its meaning. To my mind, the main property of thinking is not speed, but inertia.

Enough is not enough. In 1911,* a student of Heike Kamerlingh-Onnes noticed that at a low (4 K) temperature, mercury, which was a solid at that point, became a perfect conductor. A current would continue to flow although any driving force was absent. Kamerlingh-Onnes first thought that the student had made a mistake, that the result was nonsense. Shortly, he recognized that it was a tremendous discovery.

Kamerlingh-Onnes had studied mercury because, being a liquid at room temperature, it could be obtained in extremely pure form. He studied it at low temperatures because metals at low temperatures were known, even at that time, to have very little resistivity. Very little resistivity was expected in the experiment done by the student, but that there was absolutely no resistivity was absurd. All this happened while Niels Bohr was playing with the beginning of ideas that, fourteen years later, would lead to quantum mechanics. But even quantum mechanics could not explain superconductivity for a few decades. In the end, an explanation was produced by three American physicists, John Bardeen, Leon Cooper, and John Schrieffer, who jointly won the Nobel prize in 1972. Their work is now more familiarly known by their initials, as the "BCS theory of superconductivity."

BCS theory is incredibly complex but is undoubtedly valid. This is the general outline of their explanation. Electrons in metals are weakly coupled with small distortions of the lattice of the atoms. Two electrons with opposite spins and moving with the same velocity but in opposite directions are coupled to the lattice in precisely the

---

* ET: It is a tradition in physics to begin every discussion of superconductivity with the story of Kamerlingh-Onnes and his poor graduate student.
WT: Why do you call him "poor"?
ET: Only Kamerlingh-Onnes received the Nobel prize in 1913 for "work leading to the production of liquid helium."

same manner. For weak interactions, they will exert twice the force and cause twice the displacement in the lattice: that amounts to four times the coupling energy and this will act like an attraction between the electrons. Of course, at short distances electrons repel each other because of their similar charges. But at slightly longer distances, this weak attraction takes over. There enters at this point the truly strange part of the theory. If in a Schrödinger wave function you exchange two electrons, the sign is changed twice. Minus is changed into plus. If two electron pairs are exchanged, the wave function remains unchanged. What has happened can only be completely described in many-dimensional space, that is, by discussing the behavior of functions that depend on several variables. But the result is that the electron pairs, together with the lattice displacements that hold the pairs together, can lead to a systematic collaboration between many pairs. These collaborating pairs can form an orderly system that does not move or a similar orderly system that moves through the crystal lattice and thus carries a current. The current persists because no single electron or electron pair can break the ranks without requiring some energy. That does not mean that they are rigidly bound to the other pairs, but only that if they change, they will regret it and want to change back.

The theory explains one more important property of superconductors: from a superconductor, all magnetic fields are ejected. Or, in other words, because magnetic fields are incompatible with superconductivity, magnetic fields destroy superconductivity. The reason is connected with the pairing of the electrons. In the absence of a magnetic field, but under the influence of any static electric field— no matter how complicated—the two electrons that are moving in the opposite directions and are "spin"ning in opposite ways have precisely the same energy. This is due to the very general rule that in elementary processes a reversal of time does not change the process. The pairs, whose formation is the first step toward superconductivity, are precisely such "time-reversed" electrons. A magnetic field strong enough to exert its influence spoils this. The presence of the magnetic field destroys the pairs and, with them, the superconductivity.

Magnetic fields can penetrate a superconducting substance only by destroying its superconductivity, and this will not happen unless the magnetic field has reached a certain "critical" strength. A superconducting ring with a current flowing around it will produce a magnetic field which is captured inside the ring and which cannot escape. This is a new way by which we can provide and preserve strong magnetic fields. And this property can be used in a number of ways.

One of the ways we can use the effect is to note that the trapped magnetic field has a further interesting property: it is quantized. One can say that the number of lines of force caught by the ring cannot be changed except in very definite, small values. These are "flux quanta." By making a tiny break in the superconducting ring, one can sneak in or sneak out a magnetic force line that amounts to a single magnetic flux quantum. This can be used to make accurate measurements of magnetic fields and, even more importantly, in computers. One may make an absurd statement about a nearly absurd situation. One may count magnetic flux quanta, instead of counting beads on an abacus,* but we count the flux quanta very much faster.

All these developments seemed promising but difficult a few years ago. To use superconductivity it was necessary to go to very low temperatures and to use the results for something more than a toy or an experiment appeared expensive. But in the late 1980s, high temperature superconductivity was discovered. You should not imagine that we mean very high temperatures, such as room temperature. In fact, even now, superconductivity requires cooling almost to liquid-air temperature. But this development has brought the application of superconductivity much closer to practicality. Actually, the development came as a surprise. The detailed theory to explain the very-low-temperature superconductors did not explain these newer materials, and so the BCS theory could not predict their discovery.

---

* WT: I have heard that you had an abacus in a glass case on top of your computer with the instructions, "In case of emergency, break glass."
ET: Such was my faith in that technology–but I never broke the glass.

The older type, the low-temperature superconductors, tended to be crystals of high symmetry. And the behavior hardly depended on the orientation of the crystal. The new superconductivity always occurs in layered oxides, rather than in metals. Above the superconducting temperature, these materials are usually not insulators like most of the oxides. Neither are they good metals. They are closest to semiconductors. The electrons are filled up to a point where Brillouin zones barely overlap. One of the layers invariably is a layer whose composition, CuO_2, one copper and two oxygens, is arranged as shown in Figure 1. This layer is called a perovskite layer.* This layer consists of a simple square (or almost square) lattice of copper atoms, with an oxygen in the middle of the straight line between each pair of copper atoms. The whole substance, which is usually described as a ceramic, consists of these layers and other layers of metal oxides which do not contain, however, two oxygens for every positively charged metal ion like the perovskite layer, but one or zero. [†]

There is, as yet, no accepted explanation why these new ceramic substances show superconductivity. It is clear that this must be due to forces at least ten times stronger than those giving rise to the old-type superconductivity. It is also clear that the maximum currents attainable in the superconducting state are much lower along the z axis, perpendicular to the layered structure. In the x–y plane very much larger currents may flow. This makes it difficult to produce

---

* Named after the Russian Count, Perovsky, who was a mineralogist in the last century.

[†] WT: Now you want to give me an oxide layer with zero oxygens!

ET: This is indeed what happens if a chemist tries to be a mathematician. The famous 1-2-3 compound discovered by Chu, the first compound to become superconducting at a temperature above that of liquid nitrogen, has two CuO_2 (copper dioxide) layers, one CuO (copper oxide) layer, two BaO (barium oxygen) layers, and one Y (yttrium) layer without any oxygen. The wave functions needed for the explanation of superconductivity are probably similar in the layers containing one oxygen and no oxygen but different in the layer containing two oxygens. But this is only a guess and I state it only for the purpose of explaining why I try to talk about layers with one oxygen and no oxygen in a similar way.

Figure 1. This is the structure of a perovskite layer. In some of the new superconductors there are two, three, or four layers of this kind separated by simple layers of positively charged ions which lie over or under the "empty" spaces between four oxygen ions. The positive ions in question may be a rare-earth ion or yttrium or a calcium ion. The Cu atoms designated with an *A* form one sublattice; those with a *B,* another. This is discussed in the text.

wires out of the polycrystalline material, which is the usual form in which the new superconductors are obtained.

In crude outline, one may guess what is happening. The material usually consists of positive ions (such as copper, lanthanum, yttrium, barium, thallium, and a few others) and of negative oxygen ions, O^{--}, which hold on rather weakly to their two extra electrons. These are not metals, because the Brillouin zones tend to be completely filled or completely empty. But one may produce a semiconductor where the energy gap between filled and empty Brillouin zones is not too great. The special and characteristic perovskite layers are crowded with oxygen ions.

In the perovskite layers, the Cu ions have an odd number of electrons which tend to distribute themselves equally between the sublattices, designated A and B in Figure 1. The exchange of an electron between the two sublattices or the exchange of an electron between a perovskite layer and a neighboring layer, together with a similar exchange electron with opposite velocity and spin, may well be the reason for the pair formation in the new superconducting materials. This is analogous to what happens in usual superconductors. But in the latter, the lattice distortions are similar to those occurring in sound (e.g., expansion or contraction of the lattice), while in the perovskite layer, we have a vibration of oxygen ions approaching alternately the sublattices A and B and having a higher frequency, which may explain the higher transition temperature.

All of this is only a guess, but it illustrates how the concepts of atomic structure and quantum mechanics may interact with new and essential discoveries about peculiar materials. The time may come for custom-made materials with properties as surprising as superconductivity.

In the meantime, there remains the practical exploitation of what has already been found—if we can produce really good wire structures with the high-temperature superconductors, we could work with much stronger magnetic fields. This could reduce the size of electric motors and generators, the same toys that Faraday demonstrated in a Christmas lecture, only much more powerful and much, much smaller than generators used in industry at present.

There is another class of applications even more exciting. We have seen that electronic computers are more than a million times better than the noisy machines where I had to turn the handle just 60 years ago, literally, to grind out the numbers for my PhD thesis. The best available machines today are in the gigaflop range. (One flop is one elementary step in the computer process per second, and a giga means one billion.) By using thin films of the new superconductors, together with other semiconductor components, we might begin to flirt with petaflop machines (10^{15} flops per sec)—a million times faster still. One will be able to promptly make the machine recall anything within the Encyclopedia Britannica and much more.

Programming for such a machine will become an incredible challenge. In working on such artificial brains, our own brains will have to find new forms of mathematical thinking.

Beyond superconductors, what else can our understanding of the structure of matter bring about? Of course, I do not know. Of all the things that are impossible to plan, science is the most impossible. I know that even having understood matter, we cannot predict what next to do with it. Therefore, should we ever find out what is the concatenation* of tricks that makes life possible; this will consist of parts that we do not even dream about today.

But enough of material things. Let me now examine how Maxwell's work in codifying electricity and magnetism, coupled with quantum mechanics, has produced a technology that begins in a vacuum. I am referring to the invention and the uses of the laser.

Many consider lasers as a purely quantum phenomenon. Actually, lasers are a part of classical as well as quantum physics. Let us look first at a classical situation.

The most simple example is incident electromagnetic radiation with electric field $\mathbf{E}_i$ and an antenna which emits radiation with electric field $\mathbf{E}_a$.

We have already discussed the fact that the energy density connected with a field is proportional to the electric field squared. The energy density of the sum of electric fields $\mathbf{E}_i + \mathbf{E}_a$ is then $(\mathbf{E}_i + \mathbf{E}_a)^2$. Now I will write the incorrect equation: $(\mathbf{E}_i + \mathbf{E}_a)^2 = \mathbf{E}_i^2 + \mathbf{E}_a^2$, which says that the total energy is equal to the energy from the incident wave plus the energy from the emitted wave. It sounds reasonable, except for the fact that the equation should read: $(\mathbf{E}_i + \mathbf{E}_a)^2 = \mathbf{E}_i^2 + \mathbf{E}_a^2 + 2\mathbf{E}_i \cdot \mathbf{E}_a$ (where we recall that $\mathbf{E}_i \cdot \mathbf{E}_a$ is the magnitude of E_i times the magnitude of E_a times the cosine of the angle between them).

In most cases $2\mathbf{E}_i \cdot \mathbf{E}_a$ will be alternatingly positive and negative, depending on the frequency and the direction of the light waves. If

---

* WT: Is concatenation Hungarian?
ET: Yes, insofar as Hungarians speak Latin. Concatenation is a chain sufficiently intricate to merit a very long word.

the frequencies of the two fields are different, then we obtain positive and negative interference between the fields to the same extent and the average is zero. Even if the two frequencies agree, if I average over space I still obtain zero in most cases. It seems I can forget the term $2\mathbf{E}_i \cdot \mathbf{E}_a$.

I am, however, interested in the case where the $2\mathbf{E}_i \cdot \mathbf{E}_a$ does make a difference, where the frequencies are the same and the directions of propagation are also the same. Consider an electromagnetic wave striking an antenna that has no other power being fed into it. Now we are able to identify $\mathbf{E}_i$ as the incident electric field and recognize that it induces a current in the antenna. This current is slightly out of phase with $\mathbf{E}_i$. The current, in turn, produces $\mathbf{E}_a$, the electric field radiated by the antenna, but slightly out of phase with $\mathbf{E}_i$. In this case, the $2\mathbf{E}_i \cdot \mathbf{E}_a$ may be negative if the electric fields remain opposed, which can happen if the two waves remain in phase. The energy in the fields is *less* than you expect by $2\mathbf{E}_i \cdot \mathbf{E}_a$, because $\mathbf{E}_i$ and $\mathbf{E}_a$ tend to cancel each other. The antenna absorbs energy and the evidence of that is the shadow it casts. This agrees with common experience.

In a medium in which light absorption occurs, one may imagine the presence of many small (atomic) antennae. In each, the absorption, a decrease in intensity, is proportional to the intensity of the incident light. This results in an exponential decrease of intensity. If in, say, 10 cm the intensity is decreased by 1/2 of its original value, then in one meter it will decrease by $(1/2)^{10} = 1/1024$-fold.

What happens if the original antenna is powered, if a current runs up and down at the same frequency as that of the incident wave? In this case it is possible that behind the antenna $\mathbf{E}_i$ and $\mathbf{E}_a$ are in phase rather than out of phase. The waves reinforce each other and we have $2\mathbf{E}_i \cdot \mathbf{E}_a$ *more* energy than we expected. This is a "negative shadow"; we see additional light in the wake of the antenna. The incident light has caused the antenna to emit more light. This is called induced emission, which is the precise opposite of absorption. It is important to observe that this "negative shadow" occurs just in the region where, under normal circumstances, the shadow would appear. The added energy is found precisely "downwind" from the antenna.

Induced emission was well known to physics 50 years before it was applied. Einstein wrote a paper about it in 1917. In absorption, the decrease in light is proportional to the incident intensity. Similarly, the increase in light caused by induced emission is proportional to the light already present. This means that the light intensity grows exponentially. If the light intensity doubles in a distance of 10 cm, then in a distance of one meter the light is slightly more than 1000 times as intense. Just as our Shah found that the inventor of chess grew rich beyond any expectation, because of the exponential growth of the grains of wheat, so we find ourselves much richer in light intensity, at least, because of exponential growth in the field energy. This accounts for the miraculous properties of lasers.

All of this discussion has been based on rules valid in the macroscopic world. But quantum mechanics tells us that atoms can absorb and radiate electromagnetic energy, just as our antenna. An atom in the ground state can absorb energy—cast a shadow—and an excited atom can radiate on its own or have that radiation induced by a photon of appropriate frequency. In a laser there are more atoms (or molecules) in a certain excited state than in a state of lower energy. We call such a situation a population inversion.* A laser is no more than an assembly of atoms or molecules with a population inversion through which light propagates and is fattened up while it is propagating.

Such an arrangement might be bulky, because rather long paths may be required for the light. However, one can make lasers smaller quite easily by the use of mirrors so the light pass is folded in on itself and the population-inverted medium is used again and again.

Not only have lasers high intensity, but they can be made ex-

---

* WT: The word "inversion" refers perhaps to a situation similar to one where the student knows more than the teacher.

ET: Yes, provided that we also assume information is flowing freely from the student to the teacher.

ceedingly precise in frequency, they can be directed very accurately, and they can deliver short pulses—pulses 30 cm long easily, pulses less than a millimeter with a little effort. (These pulses last a pico-second, which is a millionth of a microsecond or 10^{-12} second.)

How do lasers actually work? Many antennae must vibrate, with exactly the same frequency, all in phase with one another. I will illuminate them. For light, the antennae are atoms or molecules and we must turn to the theory of the atom. Suppose I have an atom with its lowest level (level 1) and a higher level (level 2). If the atom is in level 1 and I illuminate it, the atom absorbs light and jumps to level 2. If the atom is in level 2, it can spontaneously emit light. This happens in a time 10^{-9} sec or more, depending on the properties of the two levels. If light falls on an atom in level 2, light will be emitted strictly in the direction to join the incident light; we have induced emission.

The phase of the vibration in level 1 automatically imitates the phase of an absorbing antenna; the phase in level 2, that of an emitting antenna. The phase of the antennae in classical electromagnetic theory is replaced in practice by the degree of occupation of the levels in the atomic theory. The correspondence between classical theory and quantum theory is a little tricky. (Exact correspondence can be obtained if atoms are replaced by harmonic oscillators.) In a population of atoms where we have as many atoms in level 1 as in level 2, then we get, on the average, neither emission nor absorption. Normally level 1 is more strongly occupied than level 2, so that absorption dominates. If you are clever, you can get a "population inversion" among the atoms. That is, there will be more atoms in level 2 than in level 1, and induced emission will dominate. So the trick of building a laser consists of producing the appropriate population inversions.

It should be remembered that a population inversion does not occur in the natural course of events, that is, it will not occur in temperature equilibrium. When we discussed statistical mechanics, we have seen that populations are determined by the Boltzmann factor $e^{-E/kT}$: the higher the energy, the smaller the population. At infinitely high temperatures, this factor becomes 1. Then the pop-

ulation is the same in an upper and a lower state. The number of
absorption and stimulated emission processes are equal and both
the absorption and the lasing effect are nil.

To bring about a population inversion, artifice is needed. A
standard trick is to use a three-level scheme, levels 1, 2, and 3 with
increasing energies. Let us assume that the energy differences are
sufficiently large so that in temperature equilibrium only level 1 is
populated. Now let us find some process that lifts the atom or molecule from level 1 to level 3. Then most of the population will still
be in 1, a smaller population will be found in 3, and practically none
will be found in 2. The result is that between 3 and 2 there is a
population inversion.

A simple and practical example is found in the case of CO_2,
carbon dioxide. Figure 2 shows the scheme of vibrational energy
levels and the arrows indicate the transitions between the levels. The
lowest level is the nonvibrating molecule. The three atoms lie on a
straight line with the carbon in the center and the two oxygens at
equal distances to the left and to the right. In level 2 there is a single
excitation of the simplest (symmetrical) normal vibration. A normal

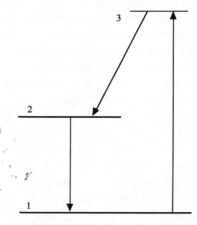

Figure 2.   Carbon dioxide energy levels permit the population inversion between
levels 3 and 2 needed for a laser.

vibration is one in which all atoms vibrate in phase. In level 3 an asymmetric vibration is excited (by one quantum) in which the oxygens move in one direction and the carbon in the opposite direction. (The carbon vibrates with a bigger amplitude, so as to keep the center of mass of the molecule unchanged.)

How shall we excite this vibration without exciting the lower 1–2 transition?

It can be done by using a peculiarity of the way molecules transfer vibrational energy to each other. As a rule, vibrations of simple molecules have a high enough frequency so that, during one collision, many vibrations occur. The result is that in collisions no vibrations are excited and no ongoing vibrations are lost (with very few exceptions). Actually, during the back-and-forth movements of a vibration, the forces of the collision partner on the vibration average out. All this is *not* true if the two colliding molecules have vibrations of quite similar frequencies. If that is the case, then vibrational energy can be transferred from one molecule to the other due to resonance.

It happens that the vibration of the N_2, nitrogen, molecule is in almost exact resonance with the 1–3 transition of the CO_2 (that is, with the asymmetric vibration). Furthermore, the N_2 vibration can be excited by a low-voltage discharge in nitrogen gas because electrons having a kinetic energy of one or two electron volts preferentially excite these nitrogen vibrations. (The reason for this phenomenon is somewhat involved and is connected with the way in which the interaction between N_2 and the electrons distort the wave function of an incident electron.) Thus by transferring energy from a discharge to the N_2 vibration and thence to level 3 of CO_2, one does produce an inversion. The 3–2 transitions give infrared light of approximately 1000 wave numbers, that is, of a wavelength of 10^{-3} cm.

But what can you do with lasers?

One thing that has been done is to measure the distance to the moon to an accuracy of one foot. (With a little care, we can make the accuracy a hundred times better, but one foot is "close enough for government work," as they say.) To do this, a corner reflector was used. A corner reflector is nothing more than three mirrors fitted

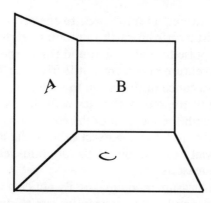

Figure 3. The corner reflector is simply three mirrors at right angles to each other. No matter from what direction light comes into this assembly, the reflected beam will go right back to the source.

together precisely perpendicular to each other to make an open corner* as in Figure 3.

A corner reflector has the nice property that it will send light back in exactly the direction from which it came. To see this, we will say that mirror A in the figure is perpendicular to the x direction, mirror B is perpendicular to the y direction, and mirror C is perpendicular to the z direction. Suppose a light beam comes into the corner reflector and hits mirror A. The x component of the light beam is reversed and the rest of the beam is bounced to, say, mirror B, where the y component of the light beam is reversed. Finally, the light beam is bounced to mirror C, where the z component is reversed. Thus the thrice reflected beam is sent back to exactly from whence it came.

A corner reflector was placed on the moon by an Apollo astronaut. An astronomer on the earth used a laser to send a very short and intense light pulse to the corner reflector. The light pulse was

---

* U.S. flyers downed at sea in World War II used corner reflectors to assist searchers in finding them.

reflected by the corner reflector back to the astronomer. He then calculated the distance from earth to moon to earth using the time it took the light pulse to make the round trip at the speed of light.

This whole venture may seem a little frivolous. For what reason do we need the distance of the moon to such great precision? After all, the distance of the moon changes with the day of the month, not to mention with the location of the earthly observation point.

We would like to know the exact orbit of the moon. This may sound like a trivial exercise, but to the doubtful reader I shall give some additional reasons.

We want to measure moonquakes. By noticing how the distance between the moon and earth changes, we can study these tremors.

We also can use the moon to study distances on earth. We can find the distance between the moon and point A on earth and the distance between the same spot on the moon and point B on earth. We will also know the angle between the lines drawn from the moon to points A and B on the earth. So with a little help from trigonometry, we can accurately find the distance between A and B.*

According to the continental drift theory, the continents of Europe and America were once connected and started to drift apart a couple of hundred million years ago. Using the moon, we can make accurate measurements of the relative positions of the continents and verify the theory.†

Lasers have applications in medicine. Glaucoma is a disease of the eye. Glaucoma causes the normal drainage of the fluids within the eye to be blocked and this causes increased pressure on the optic nerve, pressure that can eventually produce blindness. Lasers can be

---

* If we have enough pairs of points A and B, we can even determine the true shape of the earth.

† WT: I should believe this?

ET: Look at the map of the Atlantic Ocean. The bulge of Africa still fits into the Caribbean. Also (most peculiarly) the eels from Europe and America still get together each year in the mid-Atlantic at the time of the mating season. But each year they have to swim a few centimeters farther.

WT: Is their relationship platonic?

ET: I reserve the answer to our next book on biology.

used to relieve that pressure by making a small opening, an alternative to the use of drugs.

Lasers are also used in ophthalmology. Diabetes can cause the blood vessels in the eye to bleed and this can lead to blindness. Lasers can be used to cauterize these blood vessels. Lasers are about to be used to replace refractive keratoplasty, where a scalpel is used to reshape the cornea to treat deficiences like nearsightedness.

Operations where much hemorrhaging is expected are difficult or impossible to perform. Lasers can be used as the surgeon's knife, cauterizing as they cut. They also may be used with extreme accuracy. It may one day be possible to routinely operate within arteries and within the kidneys.

Communications is another field where lasers are finding new uses. The basic unit of communication is "yes" or "no." Just as in Morse code, where I can communicate any message I want with dots and dashes, so may I use another code and use a sequence of "yesses" or "noes." The test of a good communications device is how many yesses or noes I receive and can distinguish per unit time. What is a "yes" or a "no"? For a computer, we use a "1" or "0." We divide the time we are sending the message into short intervals. If a pulse is present, this means 1, and if it isn't, we have 0. By sending the message via a sharply directed beam of laser light, we can send it with the velocity of light. Even better, by using a "light pipe" made of a fiber-glass cable, we can send our message in a network within a city or even to distances beyond line of sight. Telephone companies are shifting from copper lines that carry voltage pulses to fiber-optic cables that can carry millions of telephone conversations on a few hair-thick strands of fiber glass.

The influence of lasers on chemical research has been important. One can fix the time at which an absorption process has taken place with high accuracy by using short laser pulses. One can then follow the consequences of the excitation as it influences in the interaction of molecules and chemical reactions in a precise time sequence. Such research will lead to practical applications in photochemistry.

One special case where an exact definition of frequencies is particularly important is in the separation of isotopes (that is, atoms with the same nuclear charge, the same number of electrons but different weight). The conventional methods of separation are clumsy and expensive. The difficulty is that the properties of isotopes are much too similar. But they do differ slightly in the frequencies of their spectral lines and lasers can exploit the smallest differences in frequencies.

An obvious way to proceed is to excite an isotope (or a molecule containing that isotope) and then let the excited species react. A difficulty of this method is that the reaction must occur before the excitation energy can be transferred from one atom (or molecule) to another. Unfortunately such transfers happen with great ease.

Another way is to use two (or more) quanta of different frequencies which will be absorbed in rapid succession. The first quantum is specific for its precise frequency, affecting one isotope but not another. Subsequent absorptions can occur only from the states that are already excited. The end result can be an ionized atom which can be quite rapidly separated by electromagnetic means.

One spectacular application of lasers has been discussed and pursued with great vigor: laser fusion. The idea is to concentrate a great deal of energy with the help of laser beams on a target sphere containing milligram amounts of liquid deuterium and tritium that will react with each other at high temperatures. The electric field in the laser beam is sufficiently strong to tear electrons right out of their orbits, transferring some energy to the surface of the target. This outer layer is ablated and moves swiftly outward. The remainder is forced inward, compressing the thermonuclear fuel to several hundred times liquid density.

Up to now fusion has been accomplished on earth in the hydrogen bomb. Unfortunately the explosion will not occur except in considerable amounts of nuclear fuel and this gives tremendous quantities of energy which are hard to control. If you heat a smaller amount of fuel, the arrangement will disassemble before practically any fusion can occur.

If, however, we compress a droplet of deuterium and tritium, microexplosions can be produced. A thousandfold compression permits one to reduce the amount of fuel and the explosion energy a millionfold. In this way, one may construct a "nuclear internal combustion engine."

The difficulties are mountainous: to produce the laser beams, to compress in a symmetric fashion, to handle the microexplosions (which are not all that "micro"), and to keep the whole incredible apparatus going for years.

From the point of view of the mid-21st century, all of this may appear either easy or impossible. Right now it is a part of progress into the unknown. It is probable that this will allow us to study matter in the laboratory in a state heretofore denied us—a compressed state where pressures excede those at the center of the earth.

Speaking of difficult tasks, one should mention the next step: x-ray lasers. In principle, it can be done using the spectra of highly charged ions. In practice, the job is difficult for many reasons. One of these is the extremely short lifetime of the excited states which must participate in the population inversion of an x-ray laser.

But if we should succeed, something wonderful would have been accomplished. X-ray lasers with their high intensity and their well-defined phase may be used to determine the structure of big complicated molecules which are important in biology and genetics.

That there will be practical benefits I have no doubt, but these new technologies that come from combining electromagnetic theory with quantum mechanics are certain to lead us into new realms of science.

After Maxwell constructed a comprehensive theory of electricity and magnetism, it seemed to some that physics had no new worlds to explore. The prediction was made that physics would now consist merely of determining the next figure behind the decimal point. A short time after this foolish prediction, electrons were discovered. Without these electrons neither the miracles of computers nor the follies of television would be possible. Therefore, I shall not devote any part of this text to predictions. Such, with all caution, I leave to the Epilogue, which no one should read except for the purpose of satisfying his or her curiosity.

QUESTIONS

1. By how much can k_x, k_y, and k_z of an electron in a crystal change when it participates in the emission, absorption, or scattering of light?

2. If a laser beam of 1-cm radius is emitted in the yellow part of the spectrum, what will be the minimum angular divergence of the beam? How greatly will such a beam spread when sent from earth to the moon?

3. If the planes in a corner reflector are not perpendicular to each other but deviate by one second of arc, how much will the reflected beam spread? If such an imperfect corner reflector is placed on the moon, how big an area will the reflected light cover on the earth?

4. If a dust particle contains a million molecules, how much more powerfully will it scatter a laser beam than one molecule?

5. How can a picture produced by a laser beam on a plane appear as a three-dimensional image?

EPILOGUE

After the Revolution

"Mit Eifer hab' ich mich der Studien beflissen;
Zwar weiss ich viel, doch möcht'
*ich alles wissen."**

With quantum mechanics, the explanation of the structure of matter, and the uncertainty principle, a great revolution was concluded. The last 60 years brought no truly novel intellectual developments, but there were some extremely important and widely known practical consequences. Here I shall talk mostly about theory.

Paul Adrien Maurice Dirac made the greatest contribution to a relativistic formulation of quantum mechanics. The starting point was the simple equation $E^2 - (cp)^2 = E_0^2$, where the energy E (as we have seen in Chapter 5) is the magnitude of the four component

___
* From Goethe's *Faust,* where in the first act Faust's apprentice, Wagner, says, "I studied hard. I hope I'm on the ball. Though I know much, I wish I knew it all."

energy–momentum vector, p is the three-dimensional component of this vector called the momentum, and E_0 is the minimum or rest energy.

According to Einstein, $E_0 = m_0 c^2$, where m_0 is the rest mass. But as a consequence of quantum mechanics, we have the more consistent equation, $E_0 = \pm m_0 c^2$.

What does a negative energy mean? When a particle with a positive energy and a charge e is given energy by the electric potential, its energy increases, and according to our first equation, so does its momentum. For $E_0 = -m_0 c^2$, the energy still will increase to a lesser negative value. E increases and the momentum becomes less. An electron behaving in this way has been called a donkey electron; the more you pull it, the slower it goes.

The accomplishment of Dirac was to find the proper behavior of electrons and donkey electrons. And his description included in a very elegant way the spin of the electrons as well. In the course of time, the donkey electrons with a negative energy were found in reality to be electrons with a positive charge—called positrons— (which is opposite to their normal negative charge) so that where electric fields accelerate electrons, positrons are decelerated. Instead of the energy having an unusual value, the charge has an unusual value. The positrons could also be called antielectrons. When an antielectron meets an electron, both may disappear, be annihilated, and their total energy ($2m_0 c^2$ or the difference between $+m_0 c^2$ and $-m_0 c^2$, which is $2m_0 c^2$) shows up as radiation.

As we approached the mid-century, it became clear that for every particle, there is an antiparticle. If the particle has a charge, the antiparticle has the opposite charge. Even if a particle has no charge, an antiparticle may exist. For example, in the case of the neutrino, which is a neutral particle, which in one of its forms resembles an electron, an antineutrino does exist. In the case of a light quantum, one might want to talk about an antilight quantum, but it turns out to be no different from the common light quantum. Similarly, gravitation and antigravitation turn out to be one and the same thing, which is explained by Einstein's general relativity, by the "simple" idea of space curvature.

So far, so good. From the relativistic formulae of quantum mechanics, together with the old fact that taking the square root can give positive or negative values, a consistent scheme developed. The development was crowned by the discovery of the antiproton.

A parallel and much more practical development was the experimental discovery of the neutron. Neutrons behave very similarly to protons, except for not having a charge and having slightly more mass (which means more rest energy) than protons. The proton and the neutron together turn out to be building blocks of atomic nuclei. The forces between them are of short-range (except for the electrostatic repulsion between the protons). The theory of nuclear structure remains incomplete because the nature and precise behavior of nuclear forces remains unknown.

That did not prevent a semiquantitative explanation of the behavior of nuclei. The practical results of that were the discovery of how to use nuclear fission and nuclear fusion. The first occurs in the heaviest nuclei, where repulsion between protons renders the nuclei unstable. The second is important for light nuclei and is responsible for energy production in the sun and other stars. It has been reproduced on earth on a small experimental scale as well as on a large scale.

In trying to explain the facts of nuclear physics to my very young son, I attempted to compose an atomic alphabet, which is yet to be completed, although my son is no longer young. But I did write the rhyme for the letter F:

> F stands for fission
> That is what things do
> When they get wobbly and big
> And must split in two.
> And just to complete
> The atomic confusion,
> What fission has done
> Can be undone by fusion.*

---

* WT: Can it really be undone?

ET: No. The conclusion is based on poetic license, which in this place is entirely impermissible. Fission is for the big nuclei; fusion is for the small.

Our present knowledge of nuclei is in a very rough sense similar to knowledge of the behavior of molecules during the last century. Like the chemists of that period, we have crude, practical explanations but no systematic understanding. In order to obtain such an understanding, we try to take the nucleus under a microscope, but the uncertainty principle interferes. The smaller the distance L, the bigger the uncertainty in the momentum, $\hbar/L$. Also, the energy increases as $\hbar/L$, and so does the corresponding mass $\hbar/Lc$ (remember $E = mc^2$). An appropriately small measuring rod within the nucleus is one fermi (1 F),* or 10^{-13} cm, which is a little more than 1/10 of the radius of the biggest known nuclei. The energy corresponding to one fermi is $\hbar c/L = 3 \times 10^{-4}$ g cm^2/sec, which is about 200 MeV, that is, the energy given to an electron in a potential difference of 200 million volts.

The original purpose of high-energy physics was the discovery of nuclear structure and the investigation of nuclear forces. A second purpose was to find the antiproton and that second purpose was accomplished in 1955.† The result on nuclear forces remained in-

---

* WT: I have heard that name before.
  ET: Yes. I played ping-pong with Fermi when he was starting his experiments with the newly discovered neutrons. He was also the "Italian navigator" said to have "landed safely" amidst "friendly natives" when he accomplished the first neutron chain reaction.

† ET: It was at that time, with expectations of antineutrons, antiatoms, and antimatter, that I started to speculate about the possibilities of antigalaxies. One of my friends, Harold P. Furth, thought the idea worthy of mention in *The New Yorker*:

Perils of Modern Living
by Harold Furth

Well beyond the tropostrata
There is a region stark and stellar
Where, on a streak of anti-matter,
Lived Dr. Edward Anti-Teller.

Remote from Fusion's origin,
He lived unguessed and unawares
With all his anti-kith and kin,
And kept macassars on his chairs.

One morning, idling by the sea,
He spied a tin of monstrous girth

conclusive, but instead, the first paradox made its appearance. Matter and antimatter turned out to behave differently from each other, but their similarity could be reestablished if, by going from matter to antimatter, you changed from right hand to left hand.* That followed from a closer investigation of a well-known nuclear phenomenon, β decay, in which a neutron within the nucleus turns into a proton, emitting an electron and an antineutrino.

This breaking of symmetry was followed by another observation, according to which elementary processes happen in an ever so slightly different manner when you try to reverse the process in time. Prior to this observation, future and past differed in two important ways: one the increase in disorder; the other, the unpredictability of the future. But all earlier observations maintained that each single step could be reversed, like a game of chess in which any move could be replaced by its opposite. A complete theory should explain why the mirroring of right into left and why the reversal of time fail to transform one possible change into another possible change in a precise manner.

In the meantime, higher and higher energies were employed in accelerating particles. People look inside the nucleus, at the proton and the neutron to see how they behave. We shoot neutrons and protons at each other. We shoot electrons at both of them. From these experiments we get more and more particles, all of which live a very short time. The μ meson, which has the longest life, lasts two microseconds (that is, 2×10^{-6} sec). The other particles live even a

---

That bore three letters: A.E.C.
Out stepped a visitor from Earth.

Then, shouting gladly o'er the sands,
Met two who in their alien ways
Were like as lentils. Their right hands
Clasped, and the rest was gamma rays.

The remarkable fact is that Harold got paid for the poem.

* WT: Shouldn't you have been alarmed when "Anti-Teller" seemed to offer you his left hand?
ET: Of course! Actually we should have exploded when we entered the antiatmosphere. Otherwise the adventure remains credible.

shorter time and during that short time we try to study their properties. One exception to these ephemeral particles is the neutrino, the particle with no electrical charge, zero rest mass, and travelling with the speed of light. Unfortunately for our attempts to study these creatures, they do not interact with anything except through exceedingly weak forces. A neutrino may pass through the earth with utmost nonchalance.

As new particles were discovered, there were attempts to find connections between them. Quite a few of the newcomers could be explained by assuming that the building blocks of nuclei, the neutrons and protons, are themselves composites, each of them containing three most-peculiar particles called quarks.* Their greatest peculiarity is that they seem to appear only in combinations and never alone. If they did appear singly, they would be conspicuous, because, instead of carrying the charge of an electron or multiples of electrons, they carry one-third the charge of an electron or a multiple thereof.

The attempt at systematization had a great success in unifying two quite different fields of physics: electromagnetic theory and the transformation of neutrons and protons into each other, called radioactive β decay. This established a comprehensive theory that explained the behavior of light quanta, electrons and neutrinos with the help of two high-energy particles (predicted and subsequently discovered!), carrying the designation of W and Z, and having masses almost one hundred times greater than protons and neutrons. The distance which corresponds to so high a mass is $\hbar/mc = 2 \times 10^{-16}$ cm $= 1/500$ F. The latest plans for high-energy research is to produce collisions between two particles, each carrying three hundred times the mc^2 value of those ultraheavy particles, W and Z. That would correspond to linear dimensions less than $L = 10^{-18}$ cm $= 1/100,000$ F.

The day of big physics has arrived: Accelerators encircling 1,000 square miles, expenditures on the order of $10 billion, countless scientists and engineers working on problems known to be soluble

---

* Named (with no explanation) after "three quarks for Mr. Marks," from James Joyce's *Finnegans Wake*. No statement whether Marks's quarks were ever delivered.

but requiring an enormous amount of what I hesitantly call "brute force." In my youth, I encountered a much lesser effort together with a greater content of unexpected ideas. This is a post-revolutionary period, looking for a revolution that may never come.

Can one expect new discoveries to arise in the region of ever smaller values for L? Actually, we know that L cannot be diminished below a certain value, L_{min}. To guess at such a value, we might try to compose it of the three most general constants in physics. The first is the universal gravitational constant $G = 6.7 \times 10^{-8}$ cm^3/g sec^2, which was introduced by Newton. The second is $c = 3 \times 10^{10}$ cm/sec, which Einstein recognized was the universal speed limit. The third is Planck's constant $\hbar = 1.05 \times 10^{-27}$ g cm^2/sec, which provides limits of accuracy for any observation. These three can be combined to produce a length, $L_{min} = (G\hbar/c^3)^{1/2}$ $= 1.6 \times 10^{-33}$ cm, which some call Planck's length.*

There is a clear meaning to this length. To drive microscopy beyond it is entirely and obviously impossible. This is not because, according to recent technology, it would take several billions of billions of dollars, but because of immutable rules of physics.

To overcome the gravitational attraction of the earth, a rocket must acquire a velocity of 11 km/sec. On the surface of the sun, the escape velocity is 600 km/sec. Applying our usual considerations to Planck's length, you must accept an uncertainty in momentum of $\hbar/L_{min}$ and the possibility of the presence of a mass $\hbar c/L_{min}$. To escape from this mass, concentrated in a region of dimension L_{min}, one needs 300,000 km/sec, which happens to be the speed limit c. From such a small region, containing such a mass, nothing can escape, not even information. That is called a black hole, which we mentioned in Chapter 5.

---

* ET: Note that Planck's length depends only on universal constants and not on accidental lengths such as the diameter of a proton or the wavelength of a photon of a given energy. If we take G, $\hbar$, and c, we can produce $t_{min} = L_{min}/c = 5.3 \times 10^{-44}$ sec and other interesting quantities that appear to be independent of any special physical object or condition.

WT: But could not L_{min} be called Newton's length or Einstein's length equally justifiably?
ET: You have a point, since Planck contributed only one of the three constants.

It seems that God has set the limits to human investigation far beyond our means to explore. I can understand and even share the desire of many physicists who hope that in the next step, which will carry us only to an L value one-hundred-thousandth of a fermi, we shall find the ultimate explanation of everything in physics. It would be wonderful indeed to have the ultimate answers. But I am even more attracted to the idea of a science that contains ever more surprises.

The first great revolution in physics, the Copernican revolution, led to the recognition of the existence of universal laws that can be applied both on earth and in the heavens. I do not feel that I can fully appreciate how that revolution affected the thinking and behavior of men. I have a better understanding of the effects of the second revolution, which produced a limit to the speed at which we can move in our enormously expanded universe and which turned the process of creation into one that is continuing constantly and in which each of us participates at all times. As a physicist, I hope for the continuation of surprises. Of the attributes of God, I feel most strongly about His knowing all secrets. About an everlasting God, the most appealing is that it suggests that there will be always more secrets to be discovered through never-ending surprises.

As to our present state of ignorance, I cannot as yet complain about a shortage of paradoxes. To me, the great remaining secret is that of life. I am a materialist with a difference. A materialist usually talks about matter as if he knew what he was talking about. In a way, through physics, I know that matter is something that does not quite correspond to the common understanding, but I feel I know as much about matter as a person would know about mathematics if he had just discovered how to count. The surprises that matter may yet deliver are close to infinite.

Michelangelo explained life on the ceiling of the Sistine Chapel by showing God's finger touching man. That, of course, is high art, not science. I cannot easily choose between my dissatisfaction with a divine explanation and the common materialistic explanation, which denies that there is anything to life except the accidental behavior and evolution of matter.

In the last decades, we have found out more and more about polycellular and monocellular beings, about viruses and retroviruses, about RNA and DNA, which in a highly complicated molecular form transmit the properties of parents to children. We have even obtained a crude picture of the outside of the famous double helix that is supposed to carry this information. Yet we are no closer to answering the essential question of life.

I believe that our best approach to that question will be the investigation of increasingly simple living structures. Is a virus just a poison that happens to imitate the multiplicative process of life? Or is a virus a living being that got tired of executing all the steps in the dance we call life for itself and instead is borrowing those essential processes from other living beings?

Suppose that someday we found a living being so simple that we knew every atom within it, down to the location of the last hydrogen atom. How will we find a difference between this particular molecule, which has "life," and other molecules that do not live? How will we tell, in the language of Michelangelo, that this one has been touched by the finger of God?

Niels Bohr had an answer to this question, which he would freely deliver regardless of whether or not the question was asked: "If, in a living being, you find all the details that can be determined by physical means, you will surely have killed that form of life in the process." Bohr may have been right, but the uncertainty principle became really interesting and fruitful when the conditions to which it applied were described in excruciating detail.

To me, one challenge in the inorganic world remains: to explore the frontiers between it and the living world. To arrive at those frontiers may take millennia, but it seems to me that the growth of science has become exponential. The reader of this book has, I hope, learned enough about the world of exponentials that, in it, a thousand years may become one day.

ANSWERS

CHAPTER 1

1. The missing part of the proof of the Pythagorean theorem is that we do not know that the shaded area in Fig. 2b is actually square. We know that each of the sides of the shaded figure are equal, but we have not shown that the angles are right angles. Consider the angle δ in the figure. We know that the angles $\alpha + \beta + \delta$ add up to $180°$ since they lie on a straight line. But the angles α and β are the two nonright angles in a right triangle, so $\alpha + \beta = 90°$. Therefore, δ must also be a right angle.

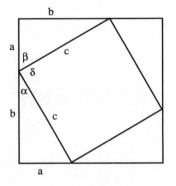

2. The velocity of light is $c = 3 \times 10^{10}$ cm/sec and the acceleration due to gravity is $g \approx 10^3$ cm/sec^2. The connection between velocity and constant acceleration* is $v = gt$, where v is the ship's speed. Then $t = c/g = 3 \times 10^7$ sec or just about one year.

3. Light traveling with the earth (or ether) either gains or loses (we assume erroneously) 3×10^6 cm/sec, the velocity of the earth about the sun. So the time to travel one meter "against" the ether will be $10^2/(3 \times 10^{10} + 3 \times 10^6)$ sec and "with" the ether $10^2/(3 \times 10^{10} - 3 \times 10^6)$ seconds. The total time "up and down" is then $10^2/(3 \times 10^{10} - 3 \times 10^6) + 10^2/(3 \times 10^{10} + 3 \times 10^6)$ sec or

$$\frac{10^{-8}}{3} \left(\frac{1}{1 - 10^{-4}} + \frac{1}{1 + 10^{-4}} \right) \text{ sec,}$$

which is equal to

$$\frac{10^{-8}}{3} [(1 + 10^{-4} + 10^{-8} + 10^{-12} + \cdots)$$

$$+ (1 - 10^{-4} + 10^{-8} - 10^{-12} + \cdots)] \text{ sec,}$$

or the total time up and down is about $(2 \times 10^{-8}/3) + (2 \times 10^{-16}/3)$ sec.

Light traveling across the "river" of ether takes the same time to go and come, so the total is

---

* If we ignore relativity, which we shouldn't.

$$\frac{2 \times 10^2}{\sqrt{(3 \times 10^{10})^2 - (3 \times 10^6)^2}} = \frac{2 \times 10^{-8}}{3\sqrt{1 - 10^{-8}}} \text{ sec.}$$

This is approximately $(2 \times 10^{-8}/3)(1 + 10^{-8}/2 + \cdots)$ sec. So we see that the light that travels across wins, but only by $10^{-16}/3$ sec. In this time, light propagates $3 \times 10^{10} \times (10^{-16}/3)$ cm $= 10^{-6}$ cm or 0.01 micron. This is 1/40 the wavelength of violet light, just sufficient to give a measurement.

CHAPTER 2

1. First I prove that three equal forces which act at 120° to each other will cancel: Suppose we rotate that configuration by 120°. The result is exactly equal to the original configuration. The only force that equals itself after a 120° rotation is the zero force, and we have proved what we wanted.

Now if I have two equal forces at 120° angles, I can add a force bisecting the angle between the forces, **c**, and so that I do not change anything; I also add an equal but opposite force **d** to the force **c**, so that it is as though **c** had not been added. But assume that forces **a**, **b**, and **d** are equal forces, making angles of 120° with each other, so they cancel. Hence **c**, which is equal to **d** in magnitude, is equal to **a** and **b** in magnitude, as well. Thus the single force that is the result of **a** and **b** is **c**, a force bisecting the angle between **a** and **b** and is equal to **a** and **b** in magnitude.

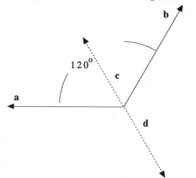

2. The water displaced by the ice cube has a weight equal to the weight of the ice cube. Thus when the ice cube melts, the resulting water will just "fill up the hole" made by the ice cube and the water level will not change.

3. The force of gravity **F**, denoted by the arrow pointed downward in the figure, can be written as the sum of two forces: **P**, pointing in the direction that the chain is supposed to travel, and the other, **Q**, perpendicular to **P**. The three forces, **F**, **P**, and **Q** form a right-angled triangle. The block on which the chain is to travel is also right angled, and you will note that the angle α of the block is equal to angle β of the triangle formed by the forces. Then the two triangles are similar and we see that the force **P** on the chain traveling along side AB will be proportional to the distance AC. Hence, for example, if AC is 1/2 of AB, then the weight of the chain on AB is twice the weight of the chain on AC, but the force **P** is only half of its weight **F**. Therefore, the force along AC will be equal to the force along AB and the system is in equilibrium.

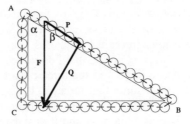

CHAPTER 3

1. The light we see from Venus is (diffusely) reflected light. When Venus is in the position shown in the figure on the left, with respect to the earth and the sun, then we will see Venus as a sickle. Now if Venus's orbit were bigger than the earth's orbit, as in the figure on the right, we will see the planet as a disc. Similarly, for any other position of the planet, we will see practically only the lighted part of the planet and it will always resemble a disc.

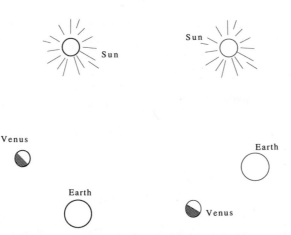

2. The apparent brightness of a star increases with its true intensity, I, and decreases as the square of the distance to the star, that is, as I/r^2. The sun appears to be 7×10^{10} times as bright as Alpha Centauri, but Alpha Centauri is actually only 20% less bright than the sun: $I_{AC} = 0.8 I_S$, so $7 \times 10^{10} = (I_S/r_S^2)/(0.8 I_S/r_{AC}^2)$ or $r_{AC}^2 = 7 \times 10^{10}(0.8) r_S^2$. For $r_S = 8$ light-minutes, this gives $r_{AC} = 19 \times 10^5$ light-minutes or 3.6 light-years.

3. If Alpha Centauri is 19×10^5 light-minutes and the orbit of earth about the sun has a radius of 8 light-minutes, then the parallax of Alpha Centauri is $8/19 \times 10^5$ radians $= 2.4 \times 10^{-4}$ degrees or $0.86''$. The astronomers say that Alpha Centauri is $1/0.86 = 1.16$ parsecs away (one parsec is about 3 light-years).

4. The new moon is not completely dark because the moon is illuminated by the sunlight scattered onto it from the "full" earth. Assuming (wrongly) that the earth scatters all the light from the sun, the intensity is reduced on its way to the moon $(R_E/D_{EM})^2$, where $R_E = 6{,}000$ km $= 0.02$ light-sec, is the radius of the earth and $D_{EM} = 360{,}000$ km $= 1.2$ light-sec, is the distance from earth to moon. Thus the new-moon to full-moon brightness would be $(R_E/D_{EM})^2 = 1/3600$.

A better expression would be $A/3600$, where A is the "albedo" or the fraction of the light scattered by the earth (A is about 0.5).

Observe that when you see the bright sickle of the moon, the rest of the moon is faintly visible. However, when the moon is almost full, you do *not* see the outline of the rest of the moon. This is because at that time the earth is nearly "new" from the moon.

5. Galileo proposed that two men stand on mountain tops at night, each with a lantern. One would open the shutter and the second would open his when he saw the light. The first man would note how long after he opened his lantern before he saw the response, the time for light to travel back and forth. Unfortunately, the mountains would be too close and human reaction time too slow for the attempt to be practical. It was over two centuries before improvements in technology allowed Fizeau to use mirrors and rotating toothed wheels as "shutters" to measure the speed of light in roughly the manner Galileo suggested.

CHAPTER 4

1. Roll the spheres down an inclined plane. The gold sphere will take 17% more time to reach the bottom. Because the densities have the ratio 7:1, the gold will have a thickness only 5% of the radius. It can then be treated as though it were a thin shell. The rotation of this shell adds a fraction $2/3$ to the inertia of forward motion, giving $1 + 2/3 = 5/3$ of inertia. The rotation of the solid sphere adds only a fraction $2/5$ to the inertia. The effective inertia will be $7/5$. The time required is proportional to the square root of the ratio of the two inertia values $\sqrt{(5/3)/(7/5)} = \sqrt{25/21} = 1.09$.

2. We know that P^2/D^3 for the apple will be equal to P^2/D^3 for the moon. The moon makes one revolution in about 27 days and it is 240,000 miles from the earth, so P^2/D^3 for the moon is $(27 \times 27)/(240,000 \times 240,000 \times 240,000) = 1/19 \times 10^{12}$. If $P = 1$ day then $1/D^3 = 1/19 \times 10^{12}$, and $D = 27,000$ miles.

CHAPTER 5

1. He compared the brightness of the supernova each night to other objects in the sky, until he finally lost it from view. By looking today at those objects, which he clearly identified, we can plot the decline in the brightness of the supernova which he observed.

2. The simple expression for the angle of deflection for light passing at distance b is obtained by taking the time integral of the component of the

acceleration parallel to b and divide by c. Thus, the increment of deflection is $GM_{sun}r^{-2}(b/r)dt/c$. Setting $r = (x^2 + b^2)^{1/2}$, we get deflection $= b/c^2 \int_{-\infty}^{\infty} GM_{sun}(x^2 + b^2)^{-3/2}dx = 2GM_{sun}/bc^2$. For b = radius of the sun, this is 0.42×10^{-5} radians.

CHAPTER 6

1. The absolute temperatures in San Francisco and Denver are 373 K (273° down to absolute zero from 0° Celsius plus 100° above 0°C) and 363 K, respectively. For any reaction, the rate, R, is given by $R = e^{-E_a/kT}$. The egg takes twice as long to cook in Denver, so R_{Den}/R_{SF} is $1/2$ and we have: $R_{Den}/R_{SF} = 1/2 = e^{-E_a/kT_{Den}}/e^{E_a/kT_{SF}} = e^{-(E_a/kT_{Den}-E_a/kT_{SF})}$. Life gets simpler if we take the logarithm of both sides before we substitute any numbers:

$$\ln 0.5 = -0.693147\cdots = -(E_a/kT_{Den} - E_a/kT_{SF})$$

$$= -(E_a/kT_{SF})(T_{SF}/T_{Den} - 1),$$

so

$$(E_a/kT_{SF}) = 0.693147/(T_{SF}/T_{Den} - 1)$$

$$= 0.693147/(1.027548 - 1) = 25.1.$$

2. The cold air from the refrigerator will furnish a small immediate relief. But in the long run, opening the refrigerator hurts. According to the laws of statistical mechanics, the cooling effect in the refrigerator is necessarily accompanied by the heating of some material. At best, you will not continue to decrease the temperature in the room. In fact, the electrical energy which is used to run the refrigerator produces heat itself so, providing that you don't care about the food in the refrigerator, the smartest thing to do would be to turn off the refrigerator.

CHAPTER 7

1. A plane metallic surface is a conducting surface. If a charge Q is placed above it, surface charges arrange themselves on the plane so that the lines

of force from the charge will intersect the plane absolutely vertically, for if they did not, currents would continue to flow across the surface of the plane. One way to visualize this is to assume two charges of equal but opposite sign at mirror image positions. All the lines coming from the one charge end up in the other. Thus we can satisfy the requirement that in electrostatics no currents will flow on the plane by imagining a second charge to be inside the plane, of equal size but opposite sign, as far below the surface as the real charge is above it, as in the figure. Of course, unlike charges attract, and the real charge tends to be pulled toward the (image charge below the) plane.

If the charge is a distance z above the plane, then the force between the two charges is $Q^2/(2z)^2$. Of course, the real charge doesn't know or care whether it is the metallic surface attracting it or the image charge—the force it feels is exactly the same.

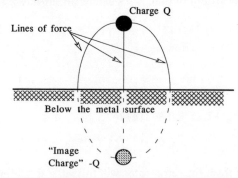

2. For convenience, take the x components of $\mathbf{E}$ and $\mathbf{H}$, derived from ϕ and $\mathbf{A}$: $E_x = (\partial A_x/\partial(ct)) - (\partial\phi/\partial x)$ and $H_x = \partial A_z/\partial y - \partial A_y/\partial z$. (We can get the y component simply by rewriting these two equations and substituting y for x, z for y, and x for z in the original. And to get the z components, we just do it again.) Now, any vector $\mathbf{A}$ that does not depend on time and any ϕ that is constant in space will give $\mathbf{E} = 0$. But what else can we say about $\mathbf{A}$ such that we will also get $\mathbf{H} = 0$ everywhere? One way is to look at $\mathbf{A}$ such that it is produced by the operation $A_x = \partial f/\partial x$, $A_y = \partial f/\partial y$, and $A_z = \partial f/\partial z$, where f can be any function of x, y, and z you choose. (The mathematicians call this operation $\mathbf{A} = \operatorname{grad} f = \nabla f$.) If you go back to the definition of the curl, you will see that if $\mathbf{A}$ is produced by this operation, its curl will always be zero, thus $\mathbf{H} = 0$.

CHAPTER 8

1. The answer can be found by dimensional reasoning. This means that we first find all the quantities on which the velocity may depend and then construct a velocity from these quantities. If this is possible only in one way, then we have obtained the way how the velocity depends on the other quantities (including the wavelength λ) that enter the problem.

Long wavelengths are governed by gravity. This, in turn, is described by the gravitational acceleration, g. The density of the liquid does not matter, because both the force and the mass are proportional to the density and the force must be divided by the mass to get the influence on the motion.

Thus only λ and g can enter. We have λ in units of cm and g in units of cm/sec^2. If we multiply the two and take the square root, we obtain $\sqrt{\lambda g}$ in units of cm/sec, a velocity. It is actually (apart from a factor $\sqrt{2}$) the velocity obtained in a fall through the distance λ. Thus velocity $\approx \sqrt{\lambda}$.

Short-wave propagation depends on surface tension σ in units of energy/cm$^2 \approx$ g cm^2 sec^{-2}/cm$^2 \approx$ g sec^{-2}. In this case the density, ρ in units of g/cm^3, is relevant. From σ, ρ, and λ, we can construct a velocity in one way only, $\sqrt{\sigma/\rho\lambda}$ in units of cm/sec and we have the wave velocity $\approx 1/\sqrt{\lambda}$.

2. In air, a good approximation is that the mean free path is 1000 molecular diameters. A molecule of air is about 2×10^{-8} cm in diameter, so the mean free path is 2×10^{-5} cm. Now we must find the average velocity at which an atom is moving. Setting the thermal energy of air is equal to its kinetic energy, so $nkT = nmv^2/2$, hence

$$v = \sqrt{2kT/m}$$
$$= \sqrt{2(1.37 \times 10^{-16})(300/28.4 \times 1.6 \times 10^{-24})}$$
$$\approx 4.25 \times 10^4 \text{ cm/sec.}$$

The time between collisions is $2 \times 10^{-5}/4.5 \times 10^4$ sec, $\approx 0.48 \times 10^{-9}$/sec or about two billion times per sec. (One billion is a better approximation.)

3. Assume that a molecule of soap weighs 100 times as much as a hydrogen atom, that is, it weighs 1.66×10^{-22} g. In a bar of soap weighing 1/4 kg $= 2.50 \times 10^2$ g, there are $2.50 \times 10^2/1.66 \times 10^{-22} = 1.5 \times 10^{24}$ soap

molecules. The diameter of a soap molecule is about 10^{-7} cm, so the area covered by one molecule is about 10^{-14} cm^2. Then the area covered by a monomolecular layer from a bar of soap would be $10^{-14} \times 1.5 \times 10^{24} = 1.5 \times 10^{10}$ cm^2 = 1.5×10^6 m^2, which is about one-half of a square mile.

CHAPTER 9

1. The simplest way to add the numbers from 1 to n is "the double bubble method." Write the numbers as in the figure. You can see that the sum of numbers in connected bubbles will always be $n + 1$. How many pairs of bubbles will you have? There will be $n/2$ of them, so all together the sum will be $(n + 1) \times (n/2)$ which is almost proportional to n^2. This method was discovered (probably not for the first time) by the German mathematician Gauss at the age of 6 when his class had to add the numbers from one to 100 for some infraction of Prussian discipline. He did not call it the double bubble method because "Double-Bubble" bubble gum didn't exist in his time.

As we mentioned in Chapter 9, the distance between energy levels of the rotating diatomic molecule is proportional to the level number. Thus we have shown that the statement in the chapter is true; the energy of rotating diatomic molecule at a particular energy level is proportional to the square of the number of that level.

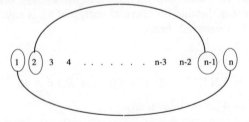

2. The centripetal acceleration of a particle moving on a circle with radius r with a velocity v is v^2/r. According to the Coulomb Law, this acceleration is equal to e^2/mr^2. Therefore, $v^2 = e^2/mr$ and the frequency $\omega = v/r$ is $\omega = (e^2/m)^{1/2}r^{-3/2}$. The kinetic energy E_{kin} is $mv^2/2$, while the potential energy, E_{pot}, is $-e^2/r$. Therefore, one sees that $E_{pot} = -2E_{kin}$ and the total energy E is $E = -e^2/2r$. This tells us that $\omega = 2^{3/2}(1/m^{1/2}e^2)(-E)^{3/2}$. We

assume quantum levels E_n with integer quantum numbers n and get $E_{n+1} - E_n = \hbar\omega = \hbar 2^{3/2}(1/m^{1/2}e^2)(-E_n)^{3/2}$. (Notice that on the right-hand side we do not know whether to use E_n or E_{n+1}. Then it is reasonable to assume $E_n \approx 1/n^2$, because $-1/(n+1)^2 + 1/n^2 = (2n+1)/(n+1)^2n^2 \approx 2/n^3 \approx 2(-E)^{3/2}$, which is the correct dependence of ω on E_n.) We determine the proportionality factor k by setting $E = -k/n^2$. Then $2k/n^3 = \hbar 2^{3/2}(1/m^{1/2}e^2)(k/n^2)^{3/2}$, which gives $k = me^4/2\hbar^2$. This agrees with the Balmer formula.

While this argument seems to hold only for circular orbits, the actual use of the formula is more general. Indeed, according to Kepler and Newton, both the energy and the period (i.e., the reciprocal frequency) of a planet depend only on the major axis of the ellipse and not the eccentricity. Therefore, the argument which we gave holds also for elliptic motion if instead of radius we use the semimajor axis.

3. As a simple model of a rotating diatomic molecule, one may consider a particle with mass m moving on a circle of radius r. Then the energy will be the kinetic energy $mv^2/2$ and the frequency will be $\omega = v/r$. Setting the difference of kinetic energies in consecutive quantum states n and $n + 1$ equal to $\hbar\omega$, one obtains

$$mv_{n+1}^2/2 - mv_n^2/2 \approx mv_n(v_{n+1} - v_n) = \hbar v_n/r.$$

Multiplying both sides with r/v_n, and noting that mrv_n is the angular momentum in the nth state, which we designate L_n, we get $mr(v_{n+1} - v_n) = L_{n+1} - L_n = \hbar$. That is, the difference of the angular momentum values in two consecutive states is Planck's constant, $\hbar$.

4. In consecutive quantum states n and $n + 1$ of the hydrogen atom, the angular momentum $L = mvr$ is different for two reasons. One is that v is different, the other is that r is different. The change $L_{n+1} - L_n$ is therefore given by

$$L_{n+1} - L_n = mr_n(v_{n+1} - v_n) + mv_n(r_{n+1} - r_n).$$

Let us set this difference equal to $\hbar$ and multiply both sides with $v_n/r = \omega$. We get

$$mv_n(v_{n+1} - v_n) + mv_n^2(r_{n+1} - r_n)/r_n = \hbar\omega.$$

The first term on the left-hand side approximates the change in kinetic energy, $1/2(mv_{n+1}^2 - mv_n^2)$, between the two states. In the second term on the left-hand side, mv_n^2/r_n is the centripetal force, while $r_{n+1} - r_n$ is the change of radius. The product of these two factors is the change in potential energy. The whole left-hand side is the change in the whole energy $E_{n+1} - E_n$, which as the equation shows, is equal to $\hbar\omega$.

Therefore, the statement that going from one circular orbit to the next one, the angular momentum changes by h leads to the same conclusion as $E_{n+1} - E_n = \hbar\omega$. It is important to notice that this holds not only for the hydrogen atom but for any situation with an axial symmetry.

One may say that the angular momentum is quantized and that the quantum is $\hbar$. Actually, circularly polarized light carries an angular momentum and Maxwell's theory shows that the angular momentum is equal to the energy of the light, divided by its frequency (divided by 2π), that is, by ω. Thus if light came in quanta of energy $\hbar\omega$, then polarized light quanta carry an angular momentum $\hbar$.

CHAPTER 10

1. By definition $\omega/k = c(1 - \alpha\omega^{-2})^{-1/2}$. Therefore, $k = c^{-1}(1 - \alpha\omega^{-2})^{1/2}\omega$ and $dk/d\omega = c^{-1}(1 - \alpha\omega^{-2})^{-1/2}$ and the group velocity has the reciprocal value $d\omega/dk = c(1 - \alpha\omega^{-2})^{1/2}$, which is smaller than the velocity of light.

2. In the fixed field of nuclei, electrons have a wave function $\phi_{R_l}(r_i)$, where r_i stands for the coordinate of the ith electron and R_l that of the lth nucleus. This function is the solution of the Schrödinger wave equation

$$-\frac{\hbar^2}{2m}\sum_i\left(\frac{\partial^2}{\partial x_i^2} + \frac{\partial^2}{\partial y_i^2} + \frac{\partial^2}{\partial z_i^2}\right)\phi_{R_l}(r_i) + [V(r_i, R_l) - E(R_l)]\phi_{R_l}(r_i) = 0.$$

The potential energy $V(r_i, R_l)$ depends on the configuration of electrons and nuclei. The eigenvalue $E(R_l)$ depends only on the configuration of the nuclei. Actually, $E(R_l)$ is the potential energy in which the nuclei move.

Writing M_l for the mass of the lth nucleus, the complete wave function is $\psi(R_l)\phi_{R_l}(r_i)$, where $\psi(R_l)$ can be obtained from the Schrödinger wave equation

$$-\sum_l\frac{\hbar^2}{2M_l}\left(\frac{\partial^2}{\partial x_l^2} + \frac{\partial^2}{\partial y_l^2} + \frac{\partial^2}{\partial z_l^2}\right)\psi(R_l) + [E(R_l) - E]\psi(R_l) = 0.$$

The function $\psi(R_l)$ usually describes the translation and rotation of the molecule, as well as its internal oscillations.

This procedure, called the Born–Oppenheimer approximation, fails to perform the differentiation of $\phi_{R_l}(r_i)$ with regard to displacements of R_l, that is, it neglects

$$\sum_l \frac{\hbar^2}{2M_l}\left[2\left(\frac{\partial \phi}{\partial x_l}\frac{\partial \psi}{\partial x_l} + \frac{\partial \phi}{\partial y_l}\frac{\partial \psi}{\partial y_l} + \frac{\partial \phi}{\partial z_l}\frac{\partial \psi}{\partial z_l}\right) + \psi\left(\frac{\partial^2 \phi}{\partial x_l^2} + \frac{\partial^2 \phi}{\partial y_l^2} + \frac{\partial^2 \phi}{\partial z_l^2}\right)\right].$$

As a rule, however, these terms are small.

CHAPTER 11

1. The knowledge for the point C is not obtained at the time of passage at C but at the later time of the passage through screen B. At this later time, knowledge of the momentum (and the velocity) will be impaired, the more so the more accurately the measurement at B is carried out.

The upshot is we have obtained knowledge from the past which we cannot utilize for predicting the future. From the *calculated* accurate position—and momentum—values at C no consequences can be drawn. These calculated values will not lead to any contradictions or to any decisions between wave or particle interpretations.

One may say that predictions are dangerous particularly for the future. If the danger involved in a prediction is not incurred, no consequence follows and the uncertainty principle is not violated.

2. Einstein wanted to find the change in energy by weighing the box, using the force $g\,\Delta m$ (and then $E = mc^2$). This force acts for a time t, so there is an inherent uncertainty in the momentum, $\Delta p = g(\Delta m)t$, giving rise to an uncertainty in position $\Delta x \geq \hbar / \Delta p$. But $g\Delta x = \Delta\phi$, where ϕ is the gravitational potential. Here Bohr used an important discovery* by Einstein: clocks run slower in high gravitational potentials as $\Delta\phi/c^2 = \Delta t/t$.

---

* In retelling that story (I heard it three times in one year) Bohr actually emphasized that point.

Bohr then pointed out that $\Delta E = c^2 \Delta m = c^2 \Delta p / gt$ and that $\Delta t = t \Delta \phi /$ $c^2 = tg \Delta x / c^2$. Thus, $\Delta E \Delta t = \Delta p \Delta x \geq \hbar$.

CHAPTER 12

1. We shall designate the wave number of light as $\kappa = 2\pi / \lambda$. Then the change of phase for light when moving by one lattice distance a will be κa $= 2\pi a / \lambda$. For visible light (that is, light carrying quanta of a few electron volts), λ is a few thousand angstroms while a is a few angstroms. Thus $2\pi a /$ $\lambda \approx 1/100$.

Taking into account that, in interation with light, we must appropriately add the phases of the lightwave to the phases of the wave functions of the electrons, the change in k (in absorption or emission) must be smaller than $1/100$. The same holds for scattering of light, which is equivalent to an absorption and an emission.

The small fraction by which k changes corresponds to the ratio of the electron velocity and the velocity of light in the material. Actually these velocities, divided into the energy, correspond to the momentum values of electrons* and light, $\hbar k / a$ and $\hbar \kappa$.

Conservation of energy conservation of momentum in an emission or absorption process will be compatible given that we jump from one Brillouin zone to another. One may say that in these processes one jumps from one Brillouin zone to another zone while k remains practically unchanged. The value of the energy and the frequency of the transition will, of course, depend on k.

2. The wavelength of yellow light is approximately 6×10^{-5} cm. When the angular deviation from the original axis is less than λ / r, where r is the radius of the beam, the light from all parts of the beam no longer gets reinforced by positive interference. In our example λ / r is 10^{-5}, that is an angle whose arc is 10^{-5} times its radius. For a second of arc, this ratio is $1/57 \times 60 \times 60$ $\approx 0.5 \times 10^{-5}$. On the way from the earth to the moon, the beam will spread by $240,000 \times 10^{-5}$ miles or 2.4 miles. Laser beams have an even smaller angular spread when produced by bigger mirrors. The small angular spread of laser beams is one of their most useful properties.

3. An error of one second of arc in a mirror produces a change of two seconds of arc in the reflected beam. This is twice the error mentioned in

---

* For electrons, k/a is not really the momentum but behaves in many respects like the momentum.

the previous example. The spread on the earth will be approximately 5 miles. The losses of intensity due to this spread and the one mentioned above can be easily tolerated due to the high intensity of lasers.

4. One million particles will be contained in a speck of dust 100 atomic diameters or 10^{-6} cm. This is less than the wavelength of light. In the scattered light there will be positive interference, giving rise to 10^6 times greater field strengths than obtained from single molecules. The intensity is enhanced 10^{12}-fold. Thus exceedingly small dust particles can be easily seen. Using well-directed lasers of short duration, one can place a laser pulse on one cubic foot on top of a chimney—a useful procedure for pollution control. I suspect that the apparatus described in H. C. Andersen's fairy tale *The Swineherd,* which could detect what the neighbor was cooking from observing his smoke must have been such a laser.

5. A common photographic image remembers the light that has fallen on one spot of a photographic plate. In a three-dimensional image the *direction* from which the light has arrived must also be remembered. This can be done by producing small local interference patterns between the light coming from the object and light coming from a fixed reference direction. This somewhat elaborate trick can be performed using the high intensities that lasers give. (The procedure, known as "holography" was described long before lasers, but had to wait until their invention to be put into common practice.)

If you look at the plate under a microscope, you will see a myriad of parallel lines—the interference patterns. By illuminating these lines from the original direction, you can apparently reconstruct the three-dimensional object and see it "floating" in space.

INDEX